Drugs and Empires

Also by James H. Mills

CANNABIS BRITANNICA: Empire, Trade and Prohibition, 1800–1928

CONFRONTING THE BODY: The Politics of Physicality in Colonial and Post-Colonial India (*with Satadru Sen*)

MADNESS, CANNABIS AND COLONIALISM: The 'Native Only' Lunatic Asylums of British India, 1857–1900

SOCCER SOUTH ASIA: Empire, Nation, Diaspora (*with Paul Dimeo*)

SUBALTERN SPORTS: Politics and Sport in South Asia

Drugs and Empires

Essays in Modern Imperialism and Intoxication, *c.* 1500–*c.* 1930

Edited by

James H. Mills

and

Patricia Barton

palgrave
macmillan

First published 2007 by
PALGRAVE MACMILLAN
Houndmills, Basingstoke, Hampshire RG21 6XS and
175 Fifth Avenue, New York, N.Y. 10010
Companies and representatives throughout the world

PALGRAVE MACMILLAN is the global academic imprint of the Palgrave Macmillan division of St. Martin's Press, LLC and of Palgrave Macmillan Ltd. Macmillan® is a registered trademark in the United States, United Kingdom and other countries. Palgrave is a registered trademark in the European Union and other countries.

ISBN-13: 978–0–230–51651–9 hardback
ISBN-10: 0–230–51651–3 hardback

This book is printed on paper suitable for recycling and made from fully managed and sustained forest sources. Logging, pulping and manufacturing processes are expected to conform to the environmental regulations of the country of origin.

A catalogue record for this book is available from the British Library.

A catalog record for this book is available from the Library of Congress.

10 9 8 7 6 5 4 3 2 1
16 15 14 13 12 11 10 09 08 07

Printed and bound in Great Britain by
Antony Rowe Ltd, Chippenham and Eastbourne

Contents

List of Figures and Tables

Figures

Table

Acknowledgements

Preparing a book of edited chapters seems to be an inexact science, so many deserve acknowledgement for their role in putting this book together. In the first place the generous funding of both the British Academy and the Wellcome Trust should be recognised, without which it would not have been possible to organise the 'Drugs and Empires' conference from which this book takes its inspiration. Their funding allowed the event to be truly international in its range of participants and explains the rich variety of papers that have ended up in this book. The ESRC should also be gratefully acknowledged for providing the Research Fellowship that allowed the time to work on assembling the collection. The conference would have been chaotic if it were not for the help afforded by colleagues in the Department of History at the University of Strathclyde, notably David Brown, and by students Raj Dhillon and Dominique Somers and all deserve thanks.

The book does not contain all the papers from the conference, and those not present in this collection should be thanked for contributions that made the event so stimulating. They include William Allison, Emdad ul Haq, Shane Blackman, Muriel Laurent, Alex Klimburg, David Anderson, Alex McKay, Paul Winther, Tim Walker, Michael Lazich and Kathleen Lodwick. The editors would also like to acknowledge the patience and professionalism of those at Palgrave, and the readers who they inveigled into providing rich comments that shaped the final content of the book.

Notes on Contributors

Patricia Barton is Research Fellow at the Centre for the Social History of Health and Healthcare, University of Strathclyde. Her publications include *The Quacks and Adulterers: Colonial South Asia's Other Drug Problem* (Forthcoming 2008).

Frank Dikötter is Professor of Modern Chinese History at SOAS, London. His publications include *Crime, Punishment and the Prison in China* (2002) and (with Laamann and Zhou) *Narcotic Culture: A History of Drugs in China* (2004). Drs Laamann and Zhou are Research Fellows in the Department of History at SOAS.

Amar Farooqui is Reader in History, Hans Raj College, University of Delhi. Among his publications are *Smuggling as Subversion: Colonialism, Indian Merchants, and the Politics of Opium* (2005) and *Colonial Forest Policy in Uttarakhand, 1890–1928* (1998).

Marc Jason Gilbert is Professor in the Department of History, North Georgia College & State University. He was co-author of *World Civilizations: The Global Experience* (2004).

Elizabeth Kelly Gray is Assistant Professor of History at Towson State University.

Chima J. Korieh is Professor of African History in the Department of History of Rowan University. His publications include (edited with G. Ugo Nwokeji) *Religion, History and Politics in Nigeria* (2005) and the forthcoming (with Femi Kolapo) *Transition and Transformation: Post-Abolition Commerce, Politics, and the Societies of the Lower Niger Basin* (2006).

Dr Lars Laamann is research fellow in the Department of History at SOAS. His publications include (with Zhou Xun and Frank Dikötter) *Narcotic Culture: A History of Drugs in China* (Chicago: Chicago University Press, 2004).

William B. McAllister is the author of *Drug Diplomacy in the Twentieth Century: An International History* (2000). He is, at the time of writing, working for the US State Department.

James H. Mills is Director of the Centre for the Social History of Health and Healthcare, Glasgow, and ESRC Research Fellow/Senior Lecturer in History, University of Strathclyde. His publications include *Madness, Cannabis and Colonialism: The 'native-only' lunatic asylums of British India, 1857–1900* (Basingstoke: Palgrave 2000) and *Cannabis Britannica: Empire, Trade and Prohibition 1800–1928* (2005).

Richard Newman is Research Associate in the Department of History at SOAS, London. His publications include 'Opium Smoking in Late Imperial China: A Reconsideration', in *Modern Asian Studies*, 29, 4, 1995.

John F. Richards is Professor in History, Duke University. His many publications include 'The Royal Commission on Opium of 1895', in *Modern Asian Studies*, 36, 2002 and 'The Opium Industry in British India', in *Indian Economic and Social History Review*, 2–3, 2002. His books include *The Unending Frontier: Environmental History of the Early Modern World* (2003) and *The Mughal Empire* (1993).

George Bryan Souza is Associate Professor in the Department of History, University of Texas, San Antonio. His publications include *The Survival of Empire: Portuguese Trade and Society in China and the South China Sea, 1630–1754* (1986) and 'Convergence before Divergence: Global Maritime Economic History and Material Culture', in *The International Journal of Maritime History*, 17, 1, 2005.

William O. Walker III is Professor of History and International Relations at the University of Toronto. Among his many publications are the books *Drug Control in the Americas* (1989) and *Opium and Foreign Policy: The Anglo-American Search for Order in Asia, 1912–1954* (1991).

Dr Xun Zhou is research fellow in the Department of History at SOAS. His publications include (with Lars Laamann and Frank Dikötter) *Narcotic Culture: A History of Drugs in China* (Chicago: Chicago University Press, 2004).

1
Introduction

James H. Mills and Patricia Barton

Imperialism and intoxicants

Drugs were at the heart of the empires of the modern period. Sales of opium in China, of alcohol in Africa, of tobacco grown in the Caribbean, of tea planted in India and of cannabis across South Asia all provided the financial resources to build administrations and the economic impetus to open new markets and to control ever wider areas. While significant economically, drugs were also central to the politics of empires; Europeans drew up treaties and fought battles in Asia in order to compete for drugs routes long before the Opium Wars of the nineteenth century; the legacy of these fashioned American relationships with the wider world, drove early development at the League of Nations, shaped the emergence of Japan as an imperial entity and of course undermined China's empire. The relationship between consciousness-altering commodities and imperialism was significant in cultural terms too. The image of the dissolute Asian addict, or the helpless African or Australasian lost in alcohol, was a recurring theme in the 'Orientalist' construction of Western superiority that legitimated empire; it was also at the centre of debates about the 'civilizing' nature of imperialism as liberals and missionaries decided that these victims were in need of salvation.

While this book contends that drugs were at the heart of imperialism, it also argues that the equation might be reversed and that it is important to explore how far imperialism has shaped the global history of drugs. The need to find products to sell to Africans and Asians in exchange for prized commodities such as slaves and tea led Europeans to become merchants of drugs and drink, and their success in this role saw societies across the world become markets for intoxicants that have endured to this day and which changed habits and tastes for ever. Imperialism shaped the politics of drugs too, not least of all in the way in which the origins of the international regulatory systems can be traced to the rivalries and diplomacy of the nineteenth- and twentieth-century empires, in which the American focus on controlling supply that still dominates was born of a desire to undermine British power in Asia. This hints at what is perhaps the greatest legacy of imperialism on the history of drugs, as it suggests that a great deal of the thinking about these consciousness-altering commodities has grown out of the cultures of colonialism. Just one example is the popular stereotype of the 'drugs epidemic', re-rehearsed periodically in discussions of intoxicants, which draws on an original portrayal of a China swept by opium addiction in the nineteenth century that grew from the exaggerations and the misconceptions of Western missionaries pursuing evangelical agendas. It seems that much of what those in the twenty-first century believe either about drugs or about imperialism draws upon the relationship between the two from *c.* 1600 into the twentieth century.

This book brings together many of the scholars working with issues of drugs and empires in order to explore a number of the assertions made above. The objective of the book is to gather together these scholars, many of whom have written book-length studies of their own on a wide range of issues and contexts, in order to encourage comparison between empires, among substances and products, and across the geographical and chronological limits that so often serve historians as organisational devices. Many of the books that deal with issues of intoxication and imperialism choose as their topic a specific substance or a certain region or territory, for the sensible reason that the detail in that subject area is itself so abundant as to demand close attention. However, this can carry the risk that wider processes or perspectives are missed.

The book is not particularly original in pursuing comparison between detailed individual studies in order to illuminate larger forces or phenomena. Previous books that have used the same device include that by William Jankowiak and Daniel Bradburd, who had published *Drugs, Labor and Colonial Expansion* in 2003. This collection draws on examples from contexts as diverse as Australia, Papua New Guinea, Namibia, Trinidad and the Andes. It focused on a particular subject, that of labour in colonial systems and the role played by psychoactive products in securing productivity in those systems. By embracing examples from across period and places, the editors reach the conclusion that 'drugs were a common feature of European expansion because their characteristics made them a particularly effective means of propagating trade or increasing the extent and intensity of labor'.[1] Here is a broad-based explanation for both consumption and supply in colonised regions.

Other stimulating collections of essays that force broader thinking on the relationship between empires and drugs include Timothy Brook and Bob Tadashi Wakabayashi's *Opium Regimes: China, Britain and Japan, 1839–1952*. By considering opiates in East Asia from a variety of perspectives alongside one another, this book moves away from the often adversarial nature of the historiography of opium in that region, which can sometimes appear to be only a thinly disguised rehearsal of the historical shifting of blame between the Chinese, the British and the Japanese empires of the period. Brook and Wakabayashi reach a conclusion that dislocates the significance of place, time or imperial entity and instead focuses on the 'vast complicities that are best understood by probing the economics, politics and cultural practices at work below the surface – not by repeating the moral claims that glossed the surface'.[2]

Consuming Habits: Drugs in History and Anthropology edited by Jordan Goodman, Paul Lovejoy and Andrew Sherratt argued for a demystification of the term 'drugs' through historical contextualisation. While their volume did not focus specifically on imperialism or imperial contexts, it did show the benefits of a comparative approach. By taking examples from societies as varied as Amerindian communities, the Muslims of Central Sudan and the Fuyuge of Papua New Guinea, they argued that socially constructed judgements of intoxicants in contemporary society can shape scholarship to the extent that 'the historical and cultural record is often distorted and poorly understood'.[3] The way forwards is simply to treat drugs as commodities, which empties them of much of their political meaning thereby allowing the historian to examine them in their own contexts. This would seem to be a particularly useful approach when dealing with psychoactive substances in colonial contexts, as these can suffer not simply from distorting judgements about drugs, but also from contemporary assumptions or controversies about the nature of imperial encounters.

Drugs and empires

If the book's primary objective is to encourage comparison between individual instances, it also serves the purpose of providing a snapshot of the way in which many of the historians who have most recently considered issues relating to drugs and empires have set about framing and approaching their subjects. It should be noted, however, that even in a book of thirteen chapters, there can be no claim to be comprehensive in terms of geographical, chronological or thematic coverage.[4] The selection of the chapters here in part simply reflects the interests of those who participated in the 'Drugs and Empires' conference in Glasgow in 2003.[5]

The book is divided into three sections which represent the key themes in the historiography on the subject of drugs history. The book deals in the first place with the issue of consumers of drugs. This reflects the recent, growing interest in what has been called the 'demand-side dynamics'[6] of markets for intoxicating products and the complex factors that shape the decision-making processes of those buying them. This interest represents a rejection among historians of the 'dominant supply-control mentality',[7] an outlook that has coloured the political search for a way of limiting drug consumption since the early twentieth century and which assumes that demand for intoxicating substances is driven by their availability. Frank Dikötter, Lars Laamann and Xun Zhou examine the market for opium in China and with it the idea that imperialist policies by the British and others created the demand for the drug. The authors argue that the market supplied by opium from India, China and elsewhere in the nineteenth and twentieth centuries must be understood in terms of China's evolving relationship with intoxicants. Smoking became embedded in a range of social activities from the sixteenth century onwards and opium became part of these from the eighteenth century onwards. It was taken in controlled contexts of respectability and conviviality, and it was important as a medicine and as an indicator of wealth and status in a period of broad economic and political change. The idea that the Chinese were unwittingly addicted by unscrupulous foreign merchants is itself a product of the period, the construction of which was aided by such diverse groups as Chinese nationalists, Western missionaries and critics of imperialism. The chapter is a call to carefully think about so many of the assumptions rehearsed in the past about the relationship between the Chinese and the opium.

George Bryan Souza explores related themes by tracing patterns of consumption of opium in early modern Asia. He uses the archives of the Dutch VOC to view its role in the supply of opium and tobacco to Southeast Asian markets in seventeenth and eighteenth centuries. The Company was important as it controlled the port at Batavia, which acted as a key entrepôt by which opium from India found its way to markets in Southeast Asia. At the same time, however, the role of the VOC was relatively limited and it was local, indigenous agents who for the most part transported opium from Batavia to Southeast Asia and beyond. Souza's chapter suggests that they did this because the demand was growing for opium alongside an increasing market for tobacco. The slow growth of smoking as a practice that spread across Asia from the sixteenth century onwards accounts for this, and Souza's chapter shows how markets for both tobacco and opium developed as consumers experimented with substances and developed complex and sophisticated tastes.

Richard Newman's chapter considers the stance taken by East India Company officials towards drugs consumers in the late eighteenth and early nineteenth century. A motley array of opinions were available to officials with which to inform themselves even before they arrived in India; they ranged from the positive evaluation of opium as a medicine to elaborate Orientalist fantasies about the drug's place in the sexual and military exploits of Asians. From the 1790s onwards, however, East India Company administrators found themselves in a position where they were expected to govern local use of intoxicants. They set about levying a tax on spirits, which yielded handsome revenues, but adopted a prohibitionist stance towards cannabis and opium products. They saw the latter as a threat to the health and sanity of the local population but the opposite proved true as Indians began to agitate for more liberal access to opium which they argued was both a medicine and a tonic. As such laws eased to make opium-eating easier and to guarantee supplies for the Indian market. However, the tone of the legislation remained focused on discouraging excessive use, albeit by relying on market mechanisms to achieve this. Newman argues that at the start of this period there was a distaste for the Indian use of intoxicants and a paternalistic sense on the part of the British that they ought to be acting to reduce levels of inebriation. It was only as a result of experience of patterns of use in subsequent decades that the British came to accept that opium-eating in moderation could be useful; quite why the Chinese would want to smoke a substance that was eaten locally was not regarded as any business of those in India.

John Richards condemns as 'presentist' the contribution of recent writers such as Emdad ul Haq who, he argues, have been influenced by current drug war rhetoric in branding as evil and criminal the impacts of colonial opium administration on India. In the first place he is critical of the idea that opium production in India encouraged a culture of opium consumption there. By the end of the nineteenth century, there was one licensed opium retailer for every 21,000 Indians, and the evidence presented at the Royal Opium Commission captures a picture of a society accustomed to consuming opium as a medicine and tonic. Indeed, Indians suspected that the attempt to squeeze opium consumption was an underhand means of driving consumers to look for alternatives in

British-produced spirits; alcohol was viewed with far greater suspicion than opium in South Asian societies, and one Indian commentator, who had traveled in the United Kingdom, went as far as to suggest that 'its substitution in place of alcohol . . . will bring back happiness to thousands of families in Great Britain and Ireland where there is no happiness now'. He also considers the economics of opium production and argues that there is little evidence that trade in the product to China had anything but a beneficial impact on the agricultural systems of areas where the plant was grown. Over a million cultivators were involved in the production of the crop in the later nineteenth century and there is no evidence of resistance on their part to the colonial system as was the case with other products such as indigo. There is little reason why there should be, as the cultivators could take advantage of interest-free loans, enjoy the security of fixed prices and find buyers for by-products from the crop. Richards concludes that here is a success story, in which the British successfully found foreign markets for Indian goods to the benefit of the local economy. If the first section considers consumers, the next section looks at the related issue of control. It does not necessarily follow that where a community consumes intoxicants a government will come along and seek to interfere in its habits. As such this section seeks to historicise state intervention in local markets and with local tastes. In doing so it traces the reasons that strategies took the shape that they did, and outlines the often unintended effects that policy could have. After all, the administrations in these chapters were not simply subject to the regular problems of modern government, but were also beset by the additional limitations and exigencies of colonial rule. The authors find ideas related to revenue, health, nationalism, Orientalism and the 'civilizing mission' among those that buffeted policy-makers where they sought to govern the predilections and addictions of the colonised.

Amar Farooqui considers the same period in India as the previous chapter but shifts the attention to Indian agents in the Malwa opium trade. In arguing that the bulk of the surpluses which accrued from opium production were appropriated by non-company participants in the enterprise in this period, he suggests that attempts by the British to control the supply of Indian opium to China were unsuccessful. In Sindia, for example, the amount of opium exported increased 12-fold in the 1820s; this provided profits for those involved in rural production, for the merchants who shipped the product, for the elites who acted as bankers and invested in the trade and for the state itself, which developed comprehensive systems for gathering duties on movements of the drug. As the British tried to tighten their grip on opium trading in the west of the region, the stakes grew higher, as profits from contraband carried with them the risk of detection. To avoid British interference, opium routes moved through Rajasthan and as far north as Sind. By the 1830s, up to 1000 chests of opium a year was making the journey to Karachi, where almost a third of customs duties were taken from opium shipments alone by the end of the decade. Even where the British did assert some sort of control over the movements of Malwa opium, it seems that the bulk of the profits remained in Indian hands. The British skimmed off a revenue by imposing a system of licensing that controlled the movement of the drug into the port city of Bombay. However, it was local merchants who bought and sold it, and as such they retained most of the spoils to be had from speculation and trade. Farooqui shows how important opium trading was to Indian agents, and argues that the British annexation of Sind and indeed the decision to embark upon the First Opium War in 1839 are closely related to the resentment of colonial administrators towards the wealth being accrued from opium by local groups in western India.

Chima Korieh's chapter also considers the colonial state's struggle to control suppliers of intoxicants, but takes as its subject the alcohol market in Nigeria. The contradictions inherent in British colonialism in Africa impacted on the liquor trade, as on the one hand commercial and administrative interests encouraged sales, while on the other moral and religious groups represented it as an evil that exploited the 'ignorant native'. The latter was a typically 'Orientalist' misrepresentation, as it failed to recognise the long local history of alcohol consumption and the well-established place of intoxicating beverages in Nigerian culture. Indeed, once local entrepreneurs began to use new technologies to produce their own brews for the market, the British authorities borrowed the language of the moralists but added a medical tone to it; the African 'gin' was represented as a toxic drink prepared by ignorant natives likely to result in the poisoning of their communities. Of course, this idiom was adopted simply in order to justify a ban on the 'gin', a measure designed to exclude a competitor to European brands from the market. The ban was ultimately unsuccessful, as Nigerians resisted colonialism by voting with their throats.

The politics of intoxicants, including alcohol, in British India is the focus of Marc Jason Gilbert's chapter. He discovers that excise duties on intoxicants represented one of the few revenue streams that the colonial state in India commanded that could be manipulated; tax on land was inelastic and tax on salt was already high by the 1870s at a time when the value of the silver rupee was decreasing. Under Lord Lytton (1876–1880) additional revenues were sought by enhancing excise collections on liquor and opium sales with success, and were more than doubled in that decade. However, the problem with this was that the colonial administration began to undermine the rhetoric of the 'civilizing mission' through its dealings in drink and drugs. Indeed, the temperance issue mingled with the anti-opium campaign both in Britain and in India to provide a potent critique of imperial rule. Even when the administration in India changed tack and successfully set out on a policy that minimised consumption by maximising revenue, critics remained hostile. In fact, many of the latter simply chose to ignore evidence that the new policy was effective in order to darken the reputation of colonial government. The agitation against colonial excise policies of the 1890s had the result, argues Gilbert, of undermining the authority and effectiveness of British imperial administration by opening up gaps between the Parliament and the Government of India. It also fuelled Indian nationalism, as the manner in which the issue was dealt with showed that both Parliamentarians in London and colonial administrators in India were unreliable and unresponsive sources of leadership for those in South Asia seeking change. From this set of conflicts over excise duties in the 1890s, temperance matters went on to become among the most potent issues championed by Gandhi to unite India's anti-colonial movement; the irony, of course, is that once in power South Asia's politicians chose to preserve the 'excise-raj' for pragmatic economic and fiscal reasons.

The effects of an imperial drugs administration in India that was so focused on intoxicants are assessed by Patricia Barton. She considers the scandal of adulterated quinine distributed by the British colonial government in the twentieth century. The reasons for this adulteration are complex, and range from the market-driven means of distributing quinine products to the poor training and payment of chemical compounders. The reason that so much of this adulteration remained undetected is more clear. Barton mentions that the purity of quinine supplies had been identified as a problem by British scientists in India as early as 1907, but that this had not excited much interest among Government of India officials. From this point onwards individual doctors pressed the issue and it was not until 1930 that an investigation was ordered by the Government of India. The chapter argues that this lack of concern was because expertise was occupied in testing in other fields; the purity and quality of opium for export was chief among the government's concerns, and the enforcement of regulations governing cannabis and cocaine later took up much of the time of the colony's Chemical Examiners and their staff. Put simply, the imperial government in India saw one of its health programmes fail because of the resources it focused on the administration of its excise policies.

The final four chapters turn from individual states and their domestic policies to the contested arena of drugs controls at the international level. In many ways the problems of colonial governments and their policies considered in the previous section explain the chapters in this final batch. By the late nineteenth century a tangle of agendas had emerged from a century or more of interference by western governments in the consumption patterns of Asians, Africans and so on. These agendas forced the issue of regulating intoxicants into international circles. Indeed, it is more accurate to argue that these agendas shaped and necessitated the emergence of the whole concept of an international approach to intoxicants and narcotics, an approach that dominates policy on such substances to this day. This third section of the book considers the origins and the forms of the international system of regulation that grew out of the period of Western imperialism.

The place of African experiences in the wider history of international drugs regulation is the focus of the chapter by James Mills. In looking at the way in which cannabis found itself entangled with the diplomatic wrangling over opium and cocaine in the 1920s at the League of Nations, Mills argues that it was African interests that forced the issue. The government of the Union of South Africa, in a letter sent in 1923, was the first to suggest that cannabis ought to be considered by the League. The reasons for this lie in the uneasy power relations of the former colony; white administrators and plantation owners had been concerned that cannabis consumption made Indian workers lazy and indolent since the 1880s, and that it encouraged fraternisation between Asian and African communities. It was the government of Egypt, however, that drove cannabis through the controversies of the 1924/1925 Geneva Opium Convention and on to the treaty agreements. The reason for this was that the Egyptian delegates enjoyed the opportunity to attack their colonial masters, the British, who were keen to leave their cannabis revenues undisturbed in India. The only evidence produced by the Egyptians, ironically, was data from the Cairo Mental Hospital which had been produced by the British superintendent there. The politics of cannabis in colonial Africa acted to thwart the British agenda at the League of Nations in 1925, and forced cannabis into the emerging regulatory framework.

The following chapter engages with conflict in the international politics of psychoactive substances in the colonial period. William Walker examines the ways in which opium consumption in China, and the trading networks that supplied the market there, became caught up in the politics of imperial rivalry between three powers in East Asia in the first decades of the twentieth century. The agenda of the British tended to be shaped by a concern for the nation's international position; as rivals industrialised successfully, the British came to believe more strongly that the empire gave it both financial and economic advantages and a superior status. India in particular needed to be shored up, and as such the significant revenues it derived from opium trading in China had to be protected. The Japanese, on the other hand, sought a foothold in Asia for economic and strategic reasons that were difficult to separate. Opiates were key products in their strategy. The United States viewed Asia from their colony in the Philippines and saw in China, a vast market for both American goods and American Christianity that was performing badly because of financial and spiritual resources squandered on opium. Finally, China itself sought to protect its political and economic integrity. Walker argues that by the mid-century, all had failed in their objectives and that none profited in the long term from engaging in opium politics.

William McAllister shows that the issue of how to deal with the trade in opium did not simply divide nations in the twentieth century, but also forced divisions within individual governments. He argues that this was the case for many of the major participants in the diplomatic processes relating to drugs in the period from the beginning of the century until the end of the Second World War, and was particularly pronounced for the British. This was reflected in the fact that the colonial government of India secured separate representation at the League of Nations meetings from the British government and actively pursued its own agenda, often in the face of the strategies of their compatriots. The factions within the British phalanx were finally forced together on to a common platform by Japanese competition in Asia and American supremacy after the Second World War. This new-found unity, however, simply reflected the extent to which the British had lost control over affairs in Asia.

The final chapter by Kelly Gray similarly focuses on divisions within, rather than between, nations on the issue of trading in drugs. Her study shows that even the American hard line on drugs is itself the outcome of complex and conflicting historical forces. She suggests that there was little public concern about the opium trade to China in the United States before the late 1830s. The drug itself was a popular medicine in the domestic market and those addicted to the substance in the United States were subjects of sympathy; regular users in China, however, were viewed as shameless Asians given to the weaknesses associated with 'other' races. American merchants involved in the opium trade, although cautious about publicising the fact, made substantial fortunes from selling the drug in China. The advent of the opium war changed attitudes. Gray shows how the conflict divided Americans, so that those with no direct interest saw the British as the villains, while those who benefited from the trade defended imperial intervention. Many Americans came to benefit from the outcome of the First Opium War, as merchants enjoyed the greater access granted to Chinese ports and missionaries seized on the opportunity to make direct contact with local communities. Nevertheless, American opinion at home remained 'newly outraged at the traffic' and a hypocrisy developed. Those who made their fortunes from opium, or whose wealth was accrued from trading in a China newly exposed by British aggression, quietly assumed positions of respectability and power once retired back to America. Gray argues that it was money from opium commerce that founded much of the modern United States; many of the antislavery schemes, hospitals, universities and railways that gave shape to nineteenth-century America would not have been possible were it not for the drugs profits made in China. The chapter is a reminder that while American attitudes to opium had hardened and cohered by the end of the nineteenth century, and went on to drive the development of the international regulatory system in the early twentieth, their position was founded on a history of ambivalence and debate.

Drugs, empires and the modern world

If the two purposes of this book already discussed are to provide snapshots of the ways in which historians have recently been investigating issues of intoxicants and imperialism, and to provide the basis for comparison and contrast, the third and final objective of this book is to stimulate debate and further research. Older historiographies in which European empires were the conscious and willing agents of addiction in non-western societies too simple and too naïve to resist the wiles of those from the west now seem unsatisfactory and indeed echoes of Orientalist fantasies and the civilising mission. That is not to dismiss modern European empires as insignificant in the history of intoxication. In the first place they were integral in the formation of a 'global economy' in which the narcotics and stimulants of specific societies were made more readily available to wider sets of consumers. This should not be overstated though, as chapters in this book by Farooqui, Souza and

Korieh, for example, remind historians that both local merchants and indigenous entrepreneurs were crucial to the development of markets for new products as a response to the opportunities created by European trading. Indeed, with an eye on the 'demand-side dynamics' mentioned above, it should be recalled that it was consumers that ultimately could decide the fate of a new product rather than suppliers.

The empires of the West were also significant as they exported the idea that control of intoxication was the business of modern government. Sometimes this took the form of state-enforced prohibition, such as that in Egypt on cannabis or in South Africa on alcohol to Africans. However, it could also range to the other extreme, of government monopolies for profit, such as that on opium in India (albeit for sale in China) or in the operation of beer halls in the twentieth-century colonial Kenya. More often than not, policy combined an element of both restriction and revenue-enhancement and could, as Newman and McAllister demonstrate in this book, fluctuate between the two. Indeed, such was the importance of intoxication to colonial regimes that it could shape domestic political agendas, as shown by Gilbert's chapter while, as Barton argues here, also having knock-on effects on unrelated state activities that compromised their outcomes. What is clear, however, is that the experience of colonialism clearly reinforced the notion that intoxication was central to the concerns and techniques of modern government, and Mills, Walker and Gray show in this book that the notion could impact not just on domestic administration, but on foreign policy too. What is equally clear is that this case must not be overstated either. Rulers in pre-colonial states across the world and throughout time have been acutely aware of both the threat and the potential of narcotics and stimulants to their regimes. Indeed, as Richard Newman demonstrates in his chapter, European colonisers often simply assumed control of pre-existing state apparatus for drawing revenue from the trade in drink and drugs. Colonial states may have devised or imported ever more complex ways for taxing or curtailing the consumption of psycho-active substances, but the principle of government interference in society's indulgences was nothing new in many parts of the colonised world.

If European empires created the mercantile routes along which intoxicants flowed, and the government systems by which access to these products was taxed and controlled, then they also formed the conditions in which ideas about such substances were formed. Gray is among those in this book who considers missionaries and religious groups, who drew on their own prejudices in condemning practices that they little understood and perceived as barriers to their own activities. Barton examines the scientists and doctors who gathered, through the fog of the nineteenth- and the early twentieth-century science, information about societies that they encountered in order to meet the rush by colonial states for data about those they were seeking to govern. State officials narrowly interpreted habits and customs through the lens of the census, the excise ledger or the policeman's gaze. Politicians, journalists and travel writers span vivid fantasies from snippets gleaned from all of the above sources, often gathered in the comfort of their metropolitan offices and shaped, as Gilbert shows, by their own interests. Empires created the need for all this information about the unfamiliar, and also the means for transmitting it quickly around the world to those that had no time to check the data, or no access to allow for verification. They also created the cultural conditions in which the misperceptions of outsiders and newcomers were privileged over the statements of members of the societies and states being scrutinised; 'Orientalist' fantasies authored by the white colonisers became authoritative knowledge about Asia, Africa, the Americas and so on. Moreover, this knowledge was then imposed both on those regions as it became the basis for policy and practice, and snuffed out much indigenous belief about psychoactive substances and intoxication.

It is evident that there is much more work to be done and a number of questions suggest themselves. Significant research has been published since the 'Drugs and Empires' conference that examines non-Western sources and which serves to explore the agendas of those in colonised societies that sought to either consume or regulate the many psychoactive substances circulating in the period of modern imperialism.[8] This has been important as it has often served to highlight the complex reasons for consumption and the many agendas of those seeking to control; the net effect has been to challenge the idea outlined above of wily Europeans peddling harmful products to simple dupes across the rest of the world whose governments were powerless to protect them. Continued research in this area promises rich results on a number of issues. More needs to be known about how far both formal and informal controls devised by pre-colonial societies informed the mechanisms adopted by colonial states, as it seems clear that few of these societies could be called a 'drug paradise', where there was unchecked access to psychoactive products. Attention also needs to be focused on the ways in which local users innovated with products, as there seems to be a continuity in ingenuity between consumers as disparate as the Chinese who added opium to their tobacco pipes in the seventeenth century and the Africans who improvised gin in the twentieth-century Nigeria. This ingenuity seems to render unlikely any effective or complete control of consumption.

Economic aspects of the relationship between drugs and empires similarly promise to be fertile ground for future investigations. 'Smuggling' and 'contraband' are slippery concepts for historians working across different periods and cultures, and yet so much of the commerce in these substances was carried on without the knowledge of, or in direct defiance of, colonial states. The challenge is to secure sources that provide glimpses of this commerce throughout the period, as recent studies of clandestine trading in opium in Western India or Central Asia have demanded reconsiderations of important topics such as the opium trade to China.[9] A focus on the economics of drugs is also important as it clouds or complicates the morally defined lines drawn between 'us' and 'them', 'consumer' and 'supplier', 'colonised' and 'coloniser' and so on, which often trouble histories of both intoxication and imperialism. After all, it is usually in the trading in psychoactive substances in the modern period where the 'complicities' are most obvious between British and Chinese, European and African, American and Colombian and so on.

The question of cultural construction in the histories both of imperialism and of intoxication is an especially important one and much remains to be done with regards to this issue. For instance, a lot has been made in this book of the changing stances taken by the Christians of the imperial era, and it would be useful to know more about how intoxicants were approached by other religions in a period when they were being transformed in meeting the cultural challenges of the West. Another rich vein of enquiry is the fate of counter-narratives in imperial nations; anyone who has read Mordecai Cooke's *Seven Sisters of Sleep* or James Lee's *Underworld of the East* will be struck by the powerful accounts of the possibilities of psychoactive substances that were gleaned from colonised societies and published by 'respectable' citizens in periods when historians have tended to focus on the condemnation and regulation of drugs. Further research on the ways in which 'science' has been created on the topics of these substances is also important, as are the processes by which this 'science' has been accepted or rejected, and by which it has survived its own historical period and has shaped or limited subsequent debates. This is especially important as so much of this 'science' was generated in colonial contexts where the practice of modern investigative techniques was fraught with Orientalist assumptions, racist beliefs and tensions with local cultures and practitioners. More also needs to be known about how knowledge about psychoactive substances was transferred and circulated, and with what effects. After all, the need for information about exotic substances and their potential effects was crucial in the moulding of scientific and medical specialisms into the academic disciplines that are now known as pharmacology, tropical medicine and so on.

When taken together, the chapters in this book serve the purpose of demanding that historians look again at the relationships between intoxication and imperialism. In doing so they will form a clearer understanding of both, and a sharper sense of the ways in which they shaped the modern world. Historians will also find themselves in a position where they can better understand the contemporary world's many confused positions on psychoactive substances, positions which, after all, have their origins in the attitudes, agreements and assumptions forged in a period when those from the West attempted to govern and dominate the globe.

Notes

1. W. Jankowiak and D. Bradburd (eds), *Drugs, Labor and Colonial Expansion* (Tucson: University of Arizona Press 2003), p. 3.
2. T. Brook and B.T. Wakabayashi (eds), *Opium Regimes: China, Britain and Japan, 1839–1952* (Berkeley: University of California Press 2000), p. 25.
3. J. Goodman, P. Lovejoy and A. Sherrat (eds), *Consuming Habits: Drugs in History and Anthropology* (London: Routledge 1995), p. 230.
4. Important books published since the *Drugs and Empires* conference in 2003 include W. Jankowiak and D. Bradburd (eds), *Drugs, Labour and Colonial Expansion*; J. Madancy, *The Troublesome Legacy of Commissioner Lin: The Opium Trade and Opium Suppression in Fujian Province, 1820s–1920s* (Cambridge: Harvard University Press 2003); J. Madancy, *Royal Commission on Opium, 1893–94: Reports, Minutes of Evidence, and Appendices* (London: Ganesha 2003); F. Dikötter, L. Laamann and Z. Xun, *Narcotic Culture: A History of Drugs in China* (London: Hurst London 2004); Z. Yangwen, *The Social Life of Opium in China* (Cambridge: Cambridge University Press 2005); D. Bello, *Opium and the Limits of Empire: Drug Prohibition in the Chinese Interior, 1729–1850* (Cambridge: Harvard University Press 2005). Recent studies that throw light on some of the many issues discussed in this book include E. Tagliacozzo, *Secret Trades, Porous Borders: Smuggling and States along a SouthEast Asian Frontier, 1865–1915* (New Haven: Yale University Press 2005); R. Mathee, *The Pursuit of Pleasure: Drugs and Stimulants in Iranian History, 1500–1900* (Princeton: Princeton University Press 2005); P. Arnaud-Chouvy and Joël Meissonnier, *Yaa Baa: Production, Traffic and consumption of Methamphetamine in Mainland South-East Asia* (Singapore: Singapore University Press 2004); E. Gebissa, *Leaf of Allah: Khat & Agricultural Transformation in Harerge, Ethiopia, 1875–1991* (Athens: Ohio University Press 2004); E. Abaka, *'Kola is God's Gift': Agricultural Production, Export Initiatives, and the Kola Industry in Asante and the Gold*

Coast, c. 1920–1950 (Athens: Ohio University Press 2005); J.C. Curto, *Enslaving Spirits: The Portuguese–Brazilian Alcohol Trade at Luanda and Its Hinterland, c. 1550–1830* (Leiden: Brill 2004); E. Kwaku Akyeampong, *Drink, Power and Cultural Change: The Social History of Alcohol in Ghana, c. 1800 to Recent Times* (Oxford: James Currey 2003); R. Moxham, *Tea: Addiction, Exploitation and Empire* (London: Constable 2003).

5. Stimulating contributions to the conference which are not included in this book include those by William Allison, Emdad ul Haq, Shane Blackman, Muriel Laurent, Alex Klimburg, David Anderson, Alex McKay, Paul Winther, Tim Walker, Michael Lazich and Kathleen Lodwick.
6. W. McAllister, *Drug Diplomacy in the Twentieth Century: An International History* (London: Routledge 2000), p. 5.
7. Ibid.
8. Chief among these are J. Madancy, *The Troublesome Legacy of Commissioner Lin*; F. Dikotter, L. Laamann and Z. Xun, *Narcotic Culture*; Z. Yangwen, *The Social Life of Opium in China*.
9. I am thinking, in particular, of A. Farooqui, *Smuggling as Subversion: Colonialism, Indian merchants and the politics of opium 1790–1843* (Oxford: Lexington 2005); D. Bello, *Opium and the Limits of Empire*.

Part I
Consumption

2

China, British Imperialism and the Myth of the 'Opium Plague'

Frank Dikötter, Lars Laamann and Xun Zhou

Introduction

In the *Cambridge History of China*, John King Fairbank, doyen of modern Chinese studies, characterised the opium trade as 'the most long-continued and systematic international crime of modern times'.[1] Indeed, in the field of modern history there appears to be a general consensus that Britain, in its merciless pursuit of financial gain, trampled on the sovereign rights of China in the early nineteenth century to enforce a shameful trade in opium. As the silver which Britain had to spend on buying tea from China began to drain the treasury, it was discovered that opium found an eager market in that country, starting a huge addiction problem, or so we are told, among the local population. Britain used overwhelming military superiority to crush the imperial army, enforce the opium traffic with gunboats, burn down the Summer Palace in Beijing and impose unequal treaties during the 'Opium Wars' of the 1840s and 1850s. In these years a highly sophisticated civilisation was powerless against the pernicious forces of an imperialist drug cartel, as Britain gradually extended its control over the ports of the country to better enforce the opium trade. The evil of opium turned China into a nation of hopeless addicts, smoking themselves to death while their country descended into chaos.

We will leave aside an examination of the complex political issues behind the 'Opium Wars' as space is limited, and will instead focus in this chapter on the belief that China was poisoned by opium.[2] The image at the core of this belief has rarely been examined, either at the time by contemporaries or more recently by historians: nationalists in China were eager to find a scapegoat in imperialism by emphasising the catastrophic results of the opium trade, at the same time as foreign missionaries and campaigning journalists published sensational reports portraying China as a victim of gunboat policy.[3] According to George T. Lay of the British and Foreign Bible Society, for instance, the typical opium smoker was characterised by 'lank and shrivelled limbs, tottering gait, sallow visage, feeble voice, and death-boding glance of the eye'.[4] The missionary T. Windsor added that opium, 'one of the devil's chief agents to bind the people to himself', kept 'hundreds of millions of people bound in absolute slavery', the only salvation being faith in Christ and belief in the Gospel.[5] During the first decades of the twentieth century, as a narcophobic discourse gradually established itself in other parts of the world, the image of China as an opium slave became the *locus classicus* of the modern drug debate, the cornerstone of the anti-opium movement and the founding case of concerted international efforts to enforce increasingly draconian measures not only against opium but against all illicit drug use in America, Europe and Asia. China is 'Patient Zero' in what is represented as a drug plague that has contaminated the rest of the globe; it is the single most important example in history of a culture commonly claimed to have been 'destroyed' by an intoxicant other than alcohol. We would like to question this image, which not only structures the entire historiography of modern China, but also underpins much of the legitimacy of today's war on drugs.

Consuming myths

The first step in dismantling the opium myth is to underline the lack of any medical evidence about the impact of opium on the health of individual consumers, bar mild constipation.[6] In nineteenth-century England, where opium was chewed and eaten in tiny portions or dissolved in tinctures by consumers of all social categories, frequent and chronic users did not suffer detrimental effects from it: many enjoyed good health well into their eighties.[7] Laudanum could be found in every home, often made to order by shopkeepers, while going to the grocer's for opium was a child's errand.[8] In South Asia, a diversity of evidence offered by both Indian and British physicians in the nineteenth century showed that opium pills were commonly taken without creating serious social or physical damage, in contrast to the strong spirits imported from abroad in the face of opposition from both the Hindu and the Muslim communities.[9]

Opium is portrayed in narcophobic discourse as a drug which produced an irresistible compulsion to increase both the amount and the frequency of dosage, although the historical evidence shows that very few users were 'compulsive addicts' who 'lost control' or suffered from a 'failure of will'. Richard Miller and others have pointed out that most reach a level beyond which they will not increase their consumption levels: users want reliable, not infinite supplies.[10] Like nicotine, opium is a psychotropic which is generally taken in determined amounts rather than ever-increasing ones: even the habitual smoker reaches a plateau, often between 7 and 15 pipes a day, a number rarely exceeded. The same daily dosage could easily be maintained year after year without developing a tolerance that required the user to smoke more and more.[11] The riddle of opium, as Jean Cocteau observed, is that the smoker never has to increase his dose.[12] Opium smokers, in short, could moderate their use for personal and social reasons and even cease taking it altogether without help. In the late 1930s, when opium prices soared in Canton, most smokers halved the amount they consumed in order to make ends meet: few would rigidly hold on to their usual dose.[13]

Another element of the opium myth is the refusal to accept that most opium use in Europe, the Middle East and Asia was light and moderate. The existence of a class of occasional, intermittent, light and moderate users was one of the most controversial issues in the opium debate in the late nineteenth century, as recognising that the majority of consumers used the substance in moderation would have undermined the case against cultivation of the poppy.[14] The denial of moderate use would also have damaged the medical argument that dosage increases could not be reversed and addiction was unavoidable, making all regular users of the drug hopeless 'addicts' and hostages to the medical authorities who alone could prevent their physical descent to certain death. However, many smokers only took up the pipe on certain occasions, for example in nineteenth-century China the official He Yongqing exclusively smoked opium to treat diarrhoea.[15] Others were intermittent smokers who drifted in and out of narcotic culture according to their personal and social requirements. Men and women would smoke a pipe or two at festivals and ceremonies several times a year without ever becoming regular users. R.A. Jamieson, a doctor in Shanghai, noted at the end of the nineteenth century that if those who smoked a few pipes on the occasion of a festival such as a marriage were to be counted, few adult males could be excluded, although habitual consumers were very rare.[16] A British consul based in Hainan also reported that 'although nearly everyone uses it . . . one never meets the opium-skeleton so vividly depicted in philanthropic works, rather the reverse – a hardy peasantry, healthy and energetic'.[17]

Another problem which needs to be addressed is the homogenisation of 'opium' into a single and uniform substance. The paste varied immensely in strength and quality, while many consumers were connoisseurs who could distinguish between a large variety of products, ranging from expensive red Persian opium to qualitatively poor local produce.[18] Opium is an extremely complex compound containing sugars, gums, acids and proteins as well as dozens of alkaloids which varied in proportion and content. General statements about the purported effects of 'opium' are thus as vague as blanket condemnations of 'alcohol': a world of difference existed between weak home-brewed beers in medieval Europe and strong spirits in Victorian England, and both were used in radically different social contexts. Most of the imported paste from India and the locally cultivated opium in China had a very low morphine content, on average 3 or 4 per cent. On the other hand, the opium imported every year into England from Turkey in tens of thousands of tonnes was very rich in morphine, ranging from 10 to 15 per cent. Moreover, smoking was generally acknowledged to be more wasteful than ingestion despite the fact that the morphine content reached the bloodstream more quickly and caused a rush: 80–90 per cent of the active compound was lost from fumes which either escaped from the pipe or were exhaled unabsorbed by the smoker.[19]

Finally, researchers working on the history of opium in China have trained their gaze exclusively on issues of supply and policy, replicating the conventional knowledge that supply determines demand.[20] However, the intricate and diverse ways in which drugs interact, collude and even collaborate with human beings in a range of diverse social contexts give psychoactive substances their particular epistemological interest.[21] Rather than focusing exclusively on the pharmacological properties of opium, it would be more fruitful for us to examine the cultural norms and social factors which sustained its consumption in the specific historical context of the late imperial period. The next section of this chapter will reconstruct the narcotic culture which endowed opium smoking with social significance in China.

Consuming culture

The spread of opium in China from the eighteenth century onwards depended on the discovery of an entirely novel mode of delivery: smoking. Our history thus starts in America, where European settlers enthusiastically adopted the local habit of inhaling tobacco before spreading it to the rest of the world. Tobacco was first introduced into

China by European traders in the late sixteenth century. The tobacco plant rapidly became a popular crop, particularly in the tropical south where it was thought to have many medicinal virtues.[22] In the hot, humid summers of the south, tobacco fumes were considered useful in fighting off miasmic diseases such as malaria; in the provinces of the north, tobacco smoking was used against the effects of cold and hunger. Yao Lü (d. 1622) was an early observer of the smoking habit and noted that 'You light one end and put the other in your mouth. The smoke goes down the throat through the pipe. It can make one tipsy, but it also protects against malaria.'[23] As tobacco was increasingly produced locally and its price became more affordable, its use spread, and it gradually became the ideal companion of tea. As such the teahouse acted as a venue for a combined activity which became known as *yancha*, namely 'smoke and tea'.[24] Guests would first smoke and then drink tea, which was supposed to cleanse the palate of the lingering taste of tobacco. Water also took time to boil, so customers were offered a smoke while waiting for tea to be prepared.[25] Tea fulfilled a variety of social roles and it took a central place in the customs of late imperial China. It became a vector of male sociability, a social lubricant, a medicinal product, a recreational item, a badge of social distinction and even a symbol of elite culture. As tea varied widely in price and quality, from rare leaves brewed with imported water down to cheap jasmine tea made with ordinary rainwater, it was also an ideal indicator of social status.

Opium appeared in the early eighteenth century laced with tobacco, a blend called *madak*.[26] The mixture was prepared by the owners of smoking houses and fetched prices significantly higher than that for pure tobacco. Opium house owners in Taiwan also provided the smoking implement, a bamboo tube with a filter made of fibres produced from local coconut palms.[27] Only by the end of the eighteenth century was the tobacco content dropped, allowing the smoking of pure opium to become a marker of social status. In a period marked by increased social mobility and conspicuous consumption, large amounts of money could be spent in one evening on pure opium. Wealth and status could be displayed far more effectively by smoking many pipes of pure opium than by drinking expensive tea or alcohol.

The shift away from *madak* towards opium was facilitated by changes in the quality of opium produced in India. Malwa opium, shipped from India by Portuguese traders, not only varied in quality but was also fiery and irritating when smoked pure. However, high-quality Patna opium, produced in India under British control, was mild and pleasant to the palate.[28] The quality of Patna further improved after poppy cultivation in Bengal was monopolised by the East India Company in 1793, the paste being bought and transported to the rest of Asia by independent traders. Affluent smokers in China appreciated the sudden improvement in the quality of opium which fuelled an ever-increasing demand for top-quality Patna in the early nineteenth century.[29]

Patna opium was an exotic commodity which became an object of connoisseurship for wealthy scholars and rich merchants during the early nineteenth century. Within these privileged circles, opium was appreciated in highly intricate and complex rituals, very much as the careful preparation of high-quality tea could confer social distinction. Terms such as 'yellow' (*huang*), 'long' (*chang*) or 'loose' (*song*) were used to describe the proper preparation of opium. A rich family normally had at least one 'opium sous-chef' to prepare the paste.[30] The cooking would be done by using two needles, one in each hand, kneading and rolling a wad of opium between the two points in the heat above the lamp. A properly trimmed wick in the lamp would generate a flame with just the right temperature, over which the carefully cooked opium would gradually acquire a dense rubbery appearance and a deep tan, its texture and colour signalling that the substance was ready for smoking. After being pulled out of the heat, a pellet was rolled into a cone and inserted into the hole of the bowl for smoking.[31]

Smoking utensils could become sought-after collectables. Expensive pipes were made of precious black wood, ivory, jade or tortoiseshell, with ornate silver decorations.[32] The stem could vary in length, the knot carved out of silver or precious wood and the bowl carefully polished. Flowers or leaves would climb along slender silver pipes, with blooming hibiscus surrounded by leaves of wild mint, while some ivory or jade pipes resembled an elephant's tusk.[33] Some connoisseurs cherished the accessories to such an extent that they became more important than the substance itself, and affluent households saw expensive pipes as a symbol of wealth and social status.[34] The close interrelation between status, consumption and connoisseurship was not confined to China, as the *chinoiserie* craze of eighteenth-century Europe shows. As tea became a sign of gentility and respectability in the higher echelons of British society, serving the beverage became associated with other novel objects of conspicuous display such as fine porcelain teaware, gilded mahogany tea furniture, silver tea *equipages* including tea caddy, teapot, tea-kettle, milk or cream jug, sugar bowl and spoon-tray.[35] China's opium utensils, likewise, were an integral part of the smoking ritual. Seduced by beautifully carved woodwork, illuminated by soft light intermingling with the smoke and the scent of opium, smokers experienced, according to enthusiasts, an intoxicating 'journey of immortality' and a veritable 'ascent to the moon' (*dao yuezhong*).[36]

In a period marked by social mobility and anxiety over class distinctions, the traditional attributes of the scholar, calligraphy, art, literature were perceived by some as being less desirable than clear markers of social status: opium was clearly one of the latter.[37] The ability to spend money on opium became a direct manifestation of wealth and status, while opium houses became known as 'money-spending holes' where customers vied to outdo each other in the conspicuous consumption of the prized substance.[38] A committed opium user would, in competitive conspicuousness, strive to become an 'opium connoisseur', connoisseurship being a carefully cultivated gentleman's art. Connoisseurs were defined not only by their expertise, but also by their ability to spend considerable amounts of money in expensive opium houses on the highest quality of opium. However, the availability of cheaper opium also meant that narcotic culture was shared by less wealthy social groups, who had little more than an oblique relationship to the cultural attributes of elite connoisseurship.

As opium smoking progressed down the social scale during the second half of the nineteenth century, it gradually became a popular marker of male sociability. It emerged as a vector of hospitality and the 'welcome smoke' (*yingchou*) offered to guests became an indispensable aspect of social etiquette with failure to offer opium considered a serious *faux pas*. Refreshments and tea would also be served, while the honoured guest reclined on a cushioned platform, at times covered in auspicious red, in order to receive his pipe.[39] Even among the less privileged, the example of the 'lonely smoker' was generally eschewed and smoking was considered a collective experience, an occasion for social intercourse, a highly ritualised event which set strict parameters for the consumption of opium. Either in opium houses or at home, opium would be smoked by friends while enjoying leisurely conversations or in groups where the pipe was passed around. During the socio-economic changes experienced in the second half of the nineteenth century, opium and teahouses as well as alcohol-serving inns provided spaces of social comfort where even less privileged groups could meet and socialise.[40]

Opium houses, contrary to the myth of the opium den as a dark and depraved trap in which the opium lamp threw a feeble light on the gaping mouths of dazed addicts, were respectable sites of male sociability in which moderate amounts of opium were shared together with tea, fruit, sweets, snacks and food. In a culture of restraint, opium was an ideal social lubricant which could be helpful in maintaining decorum and composure, in contrast to alcohol which was believed to lead to socially disruptive modes of behaviour. W. Somerset Maugham, like so many other foreign travellers in search of the mystical East, was surprised to find that the opium house he visited was neat and bright, with clean matting in every room. Far from being the expected 'dope fiends', the customers were an elderly gentleman reading his newspaper, two friends chatting over a pipe and a family with a child. The atmosphere reminded him of 'the little intimate beerhouse of Berlin where the tired working men could go in the evening and spend a peaceful hour'.[41] As late as 1930 the League of Nations reported that opium houses were often clean and tidy, failing to conform to the stereotype of 'the "opium den" as a breeding place for crime and immorality... scarcely, even at their worst, more repulsive than the localities where the corresponding classes of the Western peoples consume beer or stronger alcoholic beverages'.[42]

Rock-bottom prices at the end of the nineteenth century meant that even the very poor could participate in narcotic culture. Reports from Hangzhou described the cheapest type of opium house as open round the clock, equipped in spartan fashion with iron couches and straw mattresses, and acting as a magnet for homeless migrants and roving gamblers. Such opium houses provided many of the poor with a temporary home, bath facilities and the opportunity to eat.[43] A good example from the twentieth century is the Heng Lak Hung in Bangkok, which in the 1950s was the largest opium house in the world (opium was tolerated in Thailand until 1957). In this self-contained universe, 5000 permanent boarders enjoyed their opium, accommodation and a frugal meal, all at less than half their normal day's pay. They left in the morning for work and returned in the evening, when they stripped to underpants and slippers for a couple of hours of leisurely smoke before falling asleep on the plain wooden floors of their cubicles.[44] Opium houses in the nineteenth century fulfilled a similar role, as the treaty ports began to attract a massive influx of labourers in search of temporary work and shelter, many of whom had nowhere else to go. The decline of the rural economy, a series of natural disasters which made millions homeless and the impact of the mid-nineteenth-century rebellions forced many displaced people to seek refuge in the burgeoning cities of the coast. When the opium houses were closed in the wake of the prohibition movement in 1906, many poor people lost their homes.

Smoking was a complex social ritual with inbuilt restraints on the amount of opium which could be consumed: it had no serious consequences for the health or life expectancy of the vast majority of users. However, it was more than simply a social event or cultural device. Opium was primarily a painkiller. As Virginia Berridge noted for Victorian England, opium was a medical panacea for many working people before modern synthetic medications became available.[45] The chief motive for smoking opium in China was self-medication in order to reduce pain, fight fevers, stop diarrhoea and suppress coughs. The lowering of the cost of opium in the nineteenth century allowed ordinary people to relieve the symptoms of endemic diseases such as dysentery, cholera and malaria and to cope with fatigue, hunger and cold. Nothing was more effective than opium in treating diarrhoea.[46] 'Opium was our medicine, it was all we had', one ex-Guomindang soldier told interviewers in Thailand.[47]

The spread of affordable opium titrated in small quantities thanks to the sophisticated mechanism of the opium pipe allowed even the most dispossessed to benefit occasionally from the medical panacea in the nineteenth century. Even with the gradual spread of more modern medical facilities in the first half of the twentieth century, opium continued to be used for medical reasons by a majority of smokers. One of the few precise studies available indicates that digestive problems accounted for almost half of all the cases of medical smoking examined in Taiwan in 1907–1908, followed by diseases affecting the lungs, the nervous system and acute fevers. People living in isolated parts of the island where no medical facilities were available would regularly resort to self-medication through opium smoking.[48]

Even in the large cities of China, opium often remained the cornerstone of self-medication in the absence of effective and affordable alternatives. The North Shanghai Addiction Treatment Hospital compiled statistics for 1000 patients in 1935, of whom nearly 90 per cent cited health concerns as the main reason for taking opium or heroin. One in three took up opium smoking to relieve digestive problems, including gastric diseases and stomach pain. Coughs and spitting blood were also important reasons for reaching for the pipe; other studies mentioned stomach troubles, rheumatic pains, tuberculosis and malaria as the conditions for which opiates were most commonly taken. While poor health may have been regarded by some patients as a socially acceptable reason for smoking, the figures nonetheless reflect the overwhelming importance of opium as a painkiller, particularly in the absence of aspirin.[49] Experts in detoxification in Java acknowledged in 1930 that 80 per cent of smokers used it for medical reasons and further considered that total abolition was impractical 'because for the relief of suffering we have not yet been able to procure a substitute for opium and its derivatives'.[50]

Among the medical reasons for the popularity of opium was its imagined or genuine efficacy against the communicable diseases which destroyed countless lives in late imperial China. Fevers were endemic in the sub-tropical hills of southern China, and opium was considered an unrivalled remedy. A global epidemic of cholera, causing painful diarrhoea and bowel disorders, also reached China between 1817 and 1822.[51] Cholera arrived in the wake of opium, from India via the Straits Settlements to Canton, and further inland following established opium trading routes.[52] It may be a mere coincidence that opium and cholera proliferated in China in the same period, both were the result of a greater mobility of people and goods in an age of globalisation, but the use of opium by ordinary people was certainly encouraged by the epidemics which devastated parts of the country in the nineteenth century.

Prohibition and its consequences

This chapter so far has questioned the image of China as the victim of a 'drug plague' by explaining that opium was a culturally privileged intoxicant generally smoked in moderate amounts for recreational and medical reasons without any 'loss of control'. If smoking was a socially sanctioned practice with few adverse consequences, how could prohibition campaigns succeed in demonising it within such a short space of time? As Mike Jay has shown, while opium did not change in itself, the world around it gradually did, as the substance became a scapegoat in the politics of nationalism, a vector of racial anxieties, a bone of contention in a professional struggle against self-medication and the very foundation of a new disease theory of addiction.[53]

The second half of the nineteenth century saw the rise of the medical profession in Europe and the United States, and newly created medical associations sought moral authority and legal power by transforming opium from a folk remedy into a controlled substance. Medical associations vigorously pursued a monopoly over the opium supply, legal sanctions against opium trafficking and statutory rights to treat 'opium addicts' at a time when addiction became a growth industry for the medical profession. New notions of an 'opium plague' and a 'drug contamination' also nicely dovetailed with the moral arguments advanced by anti-opium societies in the last third of the nineteenth century: the former enabled the latter to reach a much larger audience and obtain wider public support.[54] Moreover, 'war on drugs' allowed political leaders and social elites in China to invent a fictive enemy on to whom social anxieties could be projected. Opium represented both the enemy within, the morally depraved and physically weak addict, and the enemy outside in the figure of the conniving foreign powers bent on enslaving the country. Opium became the rallying point around which social unity could be asserted, as both addicts and imperialists emerged as the ultimate alter ego against which national identity could be defined. Peddled by imperialist powers, insinuating itself into the bloodstream of the nation, poisoning the minds and bodies of the country's millions, opium gradually emerged as a symbol of national weakness, the cause of a massive 'failure of the will'. Just as the missionary became the emblematic figure of a foreign religion, opium emerged as the epitome of imperialist power and a string of anti-opium movements from 1906 onwards attempted to stamp it out.

There is abundant evidence that the transition from a tolerated opium culture to a system of prohibition from 1906 to 1949 produced a cure which was far worse than the disease. Tens of thousands of ordinary people were imprisoned and died from epidemics in crowded cells, while those deemed beyond any hope of redemption were simply executed. Opium smokers also died in detoxification centres either because the medical authorities failed effectively to treat the ailments for which opium was taken in the first place or because replacement treatments were poorly conceived and badly administered. Plenty of archival evidence exists to illustrate how opium smokers died within the first few days of treatment. One Han Zeng took advantage of an amnesty granted to volunteers for detoxification in 1915, but died on the first day of a fever: none of his relatives could be traced to collect his body.[55] In April 1946, to take an example which dates from the end of the republican era, 73-year-old Luo Bangshi, who had relied on opium to control severe gastro-intestinal problems, was compelled by the local court in Jiangsu province to follow detoxification treatment and he died in hospital on the second day of his replacement therapy.[56]

Official attempts to police the bloodstream of the nation engendered corruption, a black market and a criminal underclass. They also accelerated the spread of morphine and heroin. Both were widely smoked by a variety of social categories in the first decades of the twentieth century, although some of the heroin pills taken for recreational purposes contained only a very small proportion of alkaloids and were often based on lactose or caffeine. Morphine and heroin had few concrete drawbacks, and a number of practical advantages which persuaded many opium smokers to switch under prohibition: pills were convenient to transport, relatively cheap, odourless and thus almost undetectable in police searches, and easy to use since they no longer required the complicated paraphernalia and time-consuming rituals of opium smoking. Heroin pills, red pills in particular, enabled consumers to replicate the smoking culture created around opium while avoiding most of the problems produced by anti-opium legislation. Opium houses increasingly gave way to heroin houses, which took over many of their social functions. In Beijing, clandestine heroin houses were situated in private courtyards or secret locations advertised in pawnshops through cryptic references to dispensaries of patent medicines. A typical establishment would consist of the owner or manager, his wife, a waitress and a 'boy'. The owner normally looked after the business outside, while the wife or waitress sat by the till selling the goods. The 'boy' took care of rich patrons while keeping an eye on poor customers. He also administered injections.[57]

The link between anti-opium policies and the increased use of semi-synthetics was noted by the National Anti-Opium Association of China in 1929: 'We are quite taken by surprise by the fact that inversely as the evil practice of opium smoking is on the decrease through the united effort of the people, the extent of illicit trade in, and use of, narcotic drugs, such as morphine, heroin and cocaine, is ever on the increase.'[58] The association noted that in the provinces of Shandong and Shanxi, where opium was effectively prohibited, an unexpected expansion in the volume of morphine and heroin trading was taking place in the late 1920s. Similar observations were made by members of the association based in other provinces where new regulations were being rigidly enforced.[59] Foreign experts also noted how opium suppression encouraged heroin use. Dr K. Morinaka observed that morphine intoxication in the rest of the world was often due to dependence on or misuse of medical dugs, while in China the main motive was the search for cheaper alternatives to the opium available on the black market.[60] Government officials made similar observations. Tang Liangli, for instance, underlined that strict prohibition encouraged opium smokers to become morphine injectors. 'By enforcing drastic measures against the use of opium the Chinese Government would run the risk of increasing the number of drug addicts', Tang warned in 1935.[61]

Although some heroin pills hardly contained any alkaloids, the dirty needles shared by the poor caused lethal septicemia and transmitted a range of infectious diseases. The spread of the syringe in modern China was facilitated by a popular needle lore, which regarded the insertion of needles into the body as a therapeutic practice (acupuncture is a good example). The needle was widespread among the poor. In the 1930s, a journalist visiting a heroin house in Shanghai noted that some hypodermics were made from ordinary eye-droppers or pipettes and the more expensive ones from copper.[62] Even in local gaols, crude hand-made syringes circulated to inject shots bought for a mere ten coppers.[63] Some observers noted that in Tianjin most needles were never sterilised, often being the principal vector in the spread of syphilis. Others contracted lung infections, blood poisoning or nephritis through needle infection.[64] Wu Liande, a medical expert based in Harbin during his fight against a plague epidemic, also observed how thousands of morphine victims died every year of neglect, starvation and septicemia caused by dirty needles. In the cities of Harbin, Changchun and elsewhere in Manchuria the public health services had to bury hundreds of bodies found by the road with injection marks.[65]

Ironically, the only region where semi-synthetics failed to displace opium was the British crown colony of Hong Kong. As a result of colonial commitment to a government monopoly over the sale and distribution of opium from 1914 to 1943, the paste remained more cost-effective and convenient than heroin and morphine, and more research may show possibly that a similar stability in established patterns of intoxication characterised other colonial regimes in Asia between the two World Wars, notably Java under the Dutch Opium Regie and British India. After the colonial authorities were no longer in a position to withstand American opposition to the opium trade and were obliged to eliminate state monopolies, most opium smokers switched to heroin within less than 10 years.[66] Harold Traver suggests that the example of Hong Kong shows not only that opium prohibition generally encourages the spread of heroin, but that governments may be more successful when it comes to actively promoting drugs, whether alcohol, tobacco or opium, than in striving against all odds to prohibit them.[67]

Interestingly, the Chinese Communist Party actively participated in the lucrative opium trade during their fight against the Guomindang, and opium was one of the most important financial resources of the party, allowing it to overcome a number of fiscal difficulties and build an alternative state in the hinterland.[68] After their takeover of the country in 1949, however, it took the party a mere 3 years to eliminate all illegal substances: a dense network of police institutions, resident committees and mass organisations were used to crush drug offenders, some even being denounced by their own family members. Public trials and mass executions dealt a final blow to narcotic culture, while tens of thousands of offenders were sent to prison, often for life.[69] 'Opium', like 'prostitution', was portrayed as evil incarnate and the party spared no effort in suppressing these symbols of capitalist decadence brutally and massively.

While it is a major tenet of the opium myth that the communists 'successfully' eliminated the 'drug plague', with complete disregard for the human cost of their crusade, medical and social variables were as important in the long-term decline of narcotic culture as political factors. Penicillin appeared in the 1940s as the first antibiotic successfully to treat a whole range of diseases which had previously been managed with opiates: antibiotics took over the medical functions of opiates. On the other hand, the social status of opium was already on the decline in the 1930s, abstinence being seen as a mark of pride and a badge of modernity among social elites, very much as the rising middle classes elsewhere started to 'free' themselves from morally reprehensible 'drugs'. In Java the number of opium smokers declined drastically during the 1920s without any forceful prohibition under the Opium Regie, opium being regarded increasingly as a sign of the uncivilised while cigarette smoking took off instead.[70] Similar social differentials were also prevalent in China as opium went out of fashion. The *Shanghai Times* confidently stated in 1937 that opium-smoking was doomed in China as the practice was not 'in keeping with modern habits and outlook of life'.[71] Jean Cocteau put it more succinctly: 'Young Asia no longer smokes because "grandfathers smoked".'[72]

The cigarette, on the other hand, was fashionable and modern. Despite the spread of morphine and heroin as alternatives to opium, the commodity that most benefited from prohibition was the ready-made cigarette. It was light and palatable, easy to store and handy to use, capable of delivering nicotine straight to the lungs in short spans of time which were perfectly attuned to the faster pace of city life.[73] British American Tobacco thrived in republican China, selling half a billion cigarettes a month in a number of provinces in the 1930s,[74] and its sophisticated system of mass distribution and production was transferred to the Chinese government by Mao Zedong in 1952.[75] Tobacco cultivation and cigarette production were vigorously promoted by the CCP, as the cigarette was allowed to take over the everyday rituals and social roles of opium within a thriving smoking culture which appeared impervious to the deleterious effects of nicotine. Cigarettes evoked power and prestige and were promoted by the top leadership; Deng Xiaoping even expressed his gratitude to the cigarette as the reason for his longevity.[76] By the end of the twentieth century, China would emerge as the largest market for cigarettes and the world's leading tobacco producer.

Opium should thus be understood as part of a much wider culture of intoxication based on the inhalation of smoke. As Europe took to alcoholic and caffeinated drinks from the sixteenth century onwards, China developed a sophisticated smoking culture, starting with tobacco in the seventeenth century, followed by *madak* in the eighteenth, opium in the nineteenth and cigarettes in the twentieth. The habit of smoking crossed all social divides and included women and children, supported by the invention of ingenious smoking technologies, the opium pipe being a prime example. The advent of the cigarette in the twentieth century thus completed the smoking revolution, which had started and now ended with tobacco.

Conclusion

This chapter has suggested that opium use in China was not the outcome of aggressive imperialist policies by European powers which pushed the drug on an unsuspecting population and that did not produce a population of helpless addicts. Instead, the market supplied by opium from India, China and elsewhere[77] in the nineteenth and twentieth centuries must be understood in terms of China's evolving relationship with intoxicants. Smoking became embedded in a range of social activities from the sixteenth century onwards and opium was introduced into the rituals of consuming intoxicants associated with smoking in the eighteenth century. It was commonly taken in controlled contexts where respectability and conviviality were important, and became an important marker of wealth and status in the changing culture of the period. It was also consumed for medical reasons in an era when few effective treatments were available for even the most common ailments and diseases.

Yet the image of China as an opium slave in the nineteenth century became almost unchallengeable and was the starting point of an international war on drugs which is still being fought today. China was 'Patient Zero' in what is seen as a drug plague that has contaminated the rest of the globe, from opium in the nineteenth century to ecstasy in the twenty-first. This chapter has argued that the image was one largely constructed within discourses generated by the politics of nineteenth- and early twentieth-century imperialism. As China sought a nationalist response to the intrusions of Europeans in Asia, the stereotyped body of the opium user became both a useful symbol of China's inability to resist and an explanation of this powerlessness. At the same time, European critics of imperialism seized on Chinese critiques of the opium trade to develop a moral argument against colonial economics. This argument fused with those of modern medical professionals in the West who were keen to control access to all methods of self-medication in order to tighten their grip on the market for medical services to create an order where opium was a dangerous drug in need of control. In stimulating and justifying regimes of control that resulted in the imprisonment and execution of those who persisted in taking what had so recently been considered a useful medicine and a respectable pleasure, the anti-opium ideas of the period of high imperialism resulted in a series of cures far more dreadful than any condition created by taking opium.

Notes

1. J.K. Fairbank, 'The Creation of the Treaty System' in D. Twitchett and J.K. Fairbank (eds), *The Cambridge History of China*, volume 10 (Cambridge: Cambridge University Press 1978), p. 213.
2. A more in-depth analysis appears in F. Dikötter, L. Laamann and Xun Zhou, *Narcotic Culture: A History of Drugs in China* (London: Hurst; Chicago: University of Chicago Press 2004).
3. M. Jay, *Emperors of Dreams: Drugs in the Nineteenth Century* (Sawtry: Daedalus 2002), pp. 71–2.
4. Quoted in G.E. Morrison, *An Australian in China* (London: Cox 1895), pp. 37 and 45–6.
5. YDL, Yale Divinity School Archive, Miner Searle Bates Papers, RG 10, II, box 11, file 196, T. Windsor, 'Opium: Methods of Rescue', 1899, p. 184.
6. R.K. Newman, 'Opium Smoking in Late Imperial China: A Reconsideration' in *Modern Asian Studies*, 29, 4, 1995, pp. 767–9; another critical voice is M. Elvin, in C. Blunden and M. Elvin (eds), *Cultural Atlas of China* (Oxford: Phaidon Press 1983), pp. 148–9, as well as V.K.Y. Ho, *The City of Contrasts: Perceptions and Realities in Canton in the 1920s and 1930s* (unpublished doctoral dissertation, Oxford University 1995), pp. 79–123; J. Gray, *Rebellions and Revolutions: China from the 1800s to the 1980s* (Oxford: Oxford University Press 1990) is exemplary in his refusal to uncritically reproduce the stereotypes of dominant historiography; one of the earliest opponents of the opium myth was the sinologist H. Giles, *Some Truths about Opium* (Cambridge: Heffer 1923).
7. M. Booth, *Opium: A History* (London: Pocket Books 1997), p. 63.
8. Ibid., pp. 22–31.
9. J.F. Richards, 'Opium and the British Indian Empire: The Royal Commission of 1895' in *Modern Asian Studies*, 36, 2, 2002, pp. 406–8.
10. R.L. Miller, *The Case for Legalizing Drugs* (New York: Praeger 1991), p. 4; see also T. Goldberg, *Demystifying Drugs: A Psychosocial Perspective* (Houndsville: Macmillan 1999), pp. 22–8.
11. P. Lee, *The Big Smoke: The Chinese Art and Craft of Opium* (Bangkok: Lamplight Books 1999), p. 121.
12. J. Cocteau, *Opium: Diary of a Cure* (London: Icon 1957), p. 20.

13. V.K.Y. Ho, *'The City of Contrasts'*, p. 112.
14. V. Berridge, *Opium and the People: Opiate Use and Drug Control Policy in Nineteenth and Early Twentieth Century England* (London: Free Association Books 1999), p. 192.
15. 'Donghe zongdu Wen Chong zouqing jiang cengjing xiyapian zhi qianzong He Yongqing chigepian' (The governor general of Donghe requests the dismissal of official He Yongqing for using opium), *Junjichulu fuzouzhe*, 23 August 1840, reprinted in M. Mozhen (ed.), *Zhongguo jindu shi ziliao* (Archival Materials on the History of Drug Prohibition in China) (Tianjin: Tianjin renmin chubanshe 1998), p. 209.
16. Royal Commission on Opium, *Final Report of the Royal Commission on Opium*, volume 5 (London: Eyre and Spottiswoode 1895), p. 244.
17. Quoted in Newman, 'Opium Smoking in Late Imperial China', p. 779.
18. In his pioneering article on the history of opium in China, Jonathan D. Spence pointed out the diverse types of opium, although it is unfortunate that his otherwise exquisite piece invariably describes every smoker as an 'addict'; see J.D. Spence, 'Opium Smoking in Ch'ing China' in F. Wakeman and C. Grant (eds), *Conflict and Control in Late Imperial China* (Berkeley: University of California Press 1975), pp. 143–73.
19. J. Westermeyer, *Poppies, Pipes and People: Opium and Its Use in Laos* (Berkeley: University of California Press 1982), p. 58.
20. An excellent introduction to questions of supply and policy appears in E.R. Slack, *Opium, State, and Society: China's Narco-Economy and the Guomindang, 1924–1937* (Honolulu: University of Hawai'i Press 2000); other recent studies, in alphabetical order, include A. Baumler, *Modern China and Opium: A reader* (Michigan: University of Michigan Press 2001); A. Baumler, *Playing with Fire: The Nationalist Government and Opium in China, 1927–1941* (unpublished doctoral dissertation, University of Illinois 1997); T. Brook and B.T. Wakabayashi (eds), *Opium Regimes: China, Britain, and Japan, 1839–1952* (Berkeley: University of California Press 2000); Dong Yiming, *Étude sur le problème de l'opium dans la région du sud-ouest de la Chine (pendant les années 1920 et 1930)* (unpublished doctoral dissertation, Paris: École des Hautes Études en Sciences Sociales 1997); P.W. Howard, *Opium Suppression in Qing China: Responses to a Social Problem, 1729–1906* (unpublished doctoral dissertation, University of Pennsylvania 1998); L. Manhong, *Qingmo shehui liuxing xishi yapian yanjiu: Gongjimian zhi fenxi (1773–1906)* (A Study of Opium Consumption in Late Qing Society: A Supply-Side Analysis [1773–1906]) (unpublished doctoral dissertation, Taipei: National Normal University 1985); J. Madancy, 'Revolution, Religion and the Poppy: Opium and the Rebellion of the "Sixteenth Emperor" in Early Republican Fujian', in *Republican China*, 21, 1995, pp. 1–41; T.D. Reins, 'Reform, Nationalism and Internationalism: The Opium Suppression Movement in China and the Anglo-American Influence, 1900–1908' in *Modern Asian Studies*, 25, 1991, pp. 101–42; C.A. Trocki, *Opium, Empire and the Global Political Economy: A Study of the Asian Opium Trade 1750–1950* (London: Routledge 1999); Zhou Yongming, *Anti-Drug Crusades in Twentieth-Century China: Nationalism, History, and State-Building* (Lanham, MD: Rowman and Littlefield 1999).
21. D. Lenson, *On Drugs* (Minneapolis: University of Minnesota Press 1995), p. 60.

22. G. Cordier, *La province du Yunnan* (Hanoi: Imprimerie Mac-Dinh-Tu 1928), pp. 184–5.
23. Yao Lü, *Lushu* (A Journal of Travels and Customs), 1628–44 edition, *juan* 10, no page number.
24. Yuan Tingdong, *Zhongguo xiyan shihua* (A History of Smoking in China) (Beijing: Shangwu yinshuguan 1995), p. 48.
25. Qian Gechuan, 'Yanjiu yu pengyou' (Alcohol, Tobacco and Friends) in Lao Pin, *Mingjia bixia de yanjiuchadian* (Famous People Talking about Tobacco, Alcohol, Tea and Dim Sum) (Beijing: Zhongguo guoji guangbo chubanshe 1994), p. 341.
26. R.M. Dane, 'Historical Memorandum', in Royal Commission on Opium, *Volume VII: Final Report of the Royal Commission on Opium; Part II: Historical Appendices; Together with an Index of Witnesses and Subjects, and a Glossary of Indian Terms Used in the Evidence and Appendices. Presented to both Houses of Parliament by Command of Her Majesty* (London: Eyre and Spottiswoode 1895).
27. Huang Shujing, *Taihai shichai lu* (Journal of a Mission to Taiwan), orig. 1736 (Shanghai: Shangwu yinshuguan 1935), p. 40; Zhu Shijie, [*Xiao liuqiu manzhi*] (Account of a Voyage to Taiwan), orig. 1765 (Taipei: Datong shuju 1960), p. 54.
28. E. Watson, *The Principal Articles of Chinese Commerce* (Shanghai: Chinese Maritime Customs 1930), pp. 475–6.
29. See P. Winther, *Anglo-European Science and the Rhetoric of Empire: Malaria, Opium, and British Rule in India, 1756–1895* (Lanham: Lexington 2003), for a discussion of the Chinese preference for Bengali opium throughout the nineteenth century.
30. Xiaoliu, 'Xianzi shou' (Opium Sous-Chefs) in *Judu yuekan*, 85 (February 1935), p. 24.
31. L. Rousset, *À travers la Chine* (Paris: Hachette 1886), p. 289.
32. Ibid., p. 288.
33. See P.J. Maveety, *Opium Pipes, Prints and Paraphernalia* (Stanford: Stanford University Museum of Art 1979).
34. See the opinion of C. Farrère, *Fumée d'opium* (Paris: Mille et Une Nuits 2002), pp. 16–23.
35. J. Burnett, *Liquid Pleasures: A Social History of Drinks in Modern Britain* (London: Routledge 1999), p. 50.
36. Bing Miao, 'Yanxia zhuiyu' (Embellished Words for Smoking Clouds), *Jinyan zhuankan* (Special Publication on Opium Prohibition) (Shanghai: Shanghai jinyan weiyuanhui 1935), p. 28.
37. Zhang Changjia, 'Yanhua' (On Opium) in A Ying (ed.), *Yapian zhanzheng wenxue ji* (Collection of Literary Writings on the Opium War), volume 2 (Beijing: Guji chubanshe 1957), pp. 755–78.
38. Lu Geting, 'Chunxilu shang de "xiaojinku"' (Money spending holes on Chunxi Street) in Feng Zhicheng, *Lao Chengdu* (Old Chengdu) (Chengdu: Sichuan wenyi chubanshe 1999), pp. 133–5.
39. Rousset, *À travers la Chine*, pp. 288–9.
40. See Sun Jiazhen, *Xiaoxiang haishang fanhuameng* (The Vanity Fair of Shanghai with Portraits) (Shanghai: Shangwu yinshuguan 1915) and Sun Jiazhen, *Xu Haishang fanhuameng* (Sequel to the Vanity Fair of Shanghai) (Shanghai:

Jinbu shuju 1916); also Chen Songfeng, *Yanshi wenjianlu* (Stories on Tobacco) (Beijing: Zhongguo shangye chubanshe 1989), p. 152.

41. W.S. Maugham, *On a Chinese screen* (London: William Heinemann 1922), pp. 60–1.
42. League of Nations, C635/M254/1930/XI, Commission of Enquiry into the Control of Opium-Smoking in the Far East, *Report to the Council*, November 1930, p. 33.
43. 'Hangzhou Tongxun: Qingmo Hangzhou' (Correspondence from Hangzhou: Hangzhou During the Late Qing) in *Judu yuekan*, 98 (April 1936), p. 6.
44. R. Hughes, *Foreign Devil: Thirty Years of Reporting from the Far East* (London: Deutsch 1972), pp. 249–53.
45. Berridge, *Opium and the People*, pp. 35–8.
46. R.M. Julien, *A Primer of Drug Action* (New York: Freeman 1996), p. 247; A. Duburquois, *Notes sur les maladies des européens en Chine et au Japon* (Paris: Medical Faculty 1872), pp. 12–3.
47. J. Boyes and S. Piraban, *Opium Fields* (Bangkok: Silkworm Books 1991), p. 64.
48. K. Morinaka, 'Chronic Morphine Intoxication' in *Zhonghua yixue zazhi*, 15, 6 (December 1929), pp. 764–94.
49. Shanghai shili hubei jieyan yiyuan (ed.), *Shanghai shili hubei jieyan yiyuan nianbao fuce* (Annex to the yearly report of the Zhabei anti-opium hospital) (Shanghai: Shanghai shili hubei jieyan yiyuan 1935), pp. 2–12.
50. K.T. Sioe and T.K. Hong, 'The mass treatment of drug addiction by the Modinos' phlycten method' in Phya Damrong Baedyagun and Luang Suvejj Subhakich (eds), *Transactions of the Eighth Congress of the Far Eastern Association of Tropical Medicine* (Bangkok: Bangkok Times Press 1931), p. 53.
51. Dong Lu, Xi Ma and François Thann, *Les maux épidémiques dans l'empire chinois*, (Paris: L'Harmattan 1995), pp. 59–61; see also Yu Yunxiu, 'Liuxingxing huoluan yu Zhongguo jiu yixue' (Cholera Epidemics and Chinese Traditional Medicine), *Zhonghua yixue zazhi*, 29, 6 (December 1948), pp. 273–88.
52. J.W.H. Chun, 'An Analysis of the Cholera Problem in China with Special Reference to Shanghai' in Wu Lien-teh (ed.), *Transactions of the Ninth Congress of the Far Eastern Association of Tropical Medicine*, volume 1 (Nanjing: 1935), pp. 399–400.
53. Jay, *Emperors of Dreams*, p. 69.
54. Berridge, *Opium and the People*, pp. 193–4.
55. Beijing Muncipal Archives, J181/19/9807, 'Jiaoyangju guanyu Han Zeng jieyan yinbing siwang zhi an' (Report from the reform centre regarding the death of Han Zeng from illness whilst undergoing drug rehabilitation), 1915.
56. Jiangsu Provincial Archives, 1002/2/3473, 'Yanmin siwang ji qiangjue yanfan juan' (Files concerning cases of death and execution among drug criminals), 1946.
57. Tao Yunde, *Yapian zhi jinxi* (A History of Opium) (Shanghai: Yuzhoufeng chubanshe 1937), p. 24; Shanghai Municipal Archives, U1/4/2690, 'Report', 1935, p. 205.
58. National Anti-Opium Association of China, 'The Narcotic Situation in China' in *Zhonghua yixue zazhi*, 15, 2 (April 1929), p. 270.
59. Ibid., p. 271.

60. K. Morinaka, 'Chronic Morphine Intoxication', *Zhonghua yixue zazhi*, 15, 6 (December 1929), pp. 766, 771.
61. T'ang Leang-li, *Reconstruction in China* (Shanghai: China United Press 1935), pp. 120–1.
62. 'Heroin Injection Horrors', *North China Herald*, 8 July 1936, p. 82.
63. L.S. Huizenga, 'Autogenous Serum Treatment for Opium Addicts' in *Chinese Medical Journal*, 49, 7 (July 1935), pp. 719–21.
64. Zhonghua quanguo jidujiao xiehui judu weiyuanhui, *Duji wenda* (Questions and Answers Regarding Narcotics) (Shanghai: Zhonghua guomin juduhui 1925), p. 11.
65. Wu Lien-teh, 'Public Health Aspects of the Narcotic Problem' in *Zhonghua yixue zazhi*, 11, 6 (December 1925), p. 423.
66. On the role of the United States in pushing a policy of strict prohibition at all cost, see D.R. Bewley-Taylor, *The United States and International Drug Control, 1909–1997* (London: Pinter 1999).
67. H. Traver, 'Opium to Heroin: Restrictive Legislation and the Rise of Heroin Consumption in Hong Kong' in *Journal of Policy History*, 4, 3, 1992, pp. 320–1.
68. Ch'en Yung-fa, 'The Blooming Poppy Under the Red Sun: The Yan'an Way and the Opium Trade' in T. Saich and H. van de Ven (eds), *New perspectives on the Chinese communist revolution* (New York: Sharpe 1995), pp. 263–98; his findings are confirmed in National Archives Number Two, 41/78, 'Neizhengbu Hebei ji Ping Jin shi 1946–1948 niandu yandu dujin gongzuo jihua baogao' (Annual reports on opium suppression in 1946–1948 from the Beijing and Tianjin divisions of the Ministry of Internal Affairs),1946–1948, pp. 26–42 and attached report.
69. See various contributions in M. Weigang (ed.), *Jinchang jindu* (Eliminate Prostitution and Eradicate Opium) (Beijing: Jingguan jiaoyu chubanshe 1993); see also Zhou Yongming, *Anti-Drug Crusades in Twentieth-Century China: Nationalism, History, and State-Building* (Lanham: Rowman and Littlefield 1999).
70. J.R. Rush, 'Opium in Java: A Sinister Friend' in *Journal of Asian Studies*, 44, 3, 1985, p. 557.
71. 'The Opium Question', *The Shanghai Times*, 9 July 1937.
72. Cocteau, *Opium*, p. 18.
73. D.T. Courtwright, *Forces of Habit: Drugs and the Making of the Modern World* (Cambridge, MA: Harvard University Press 2001), p. 114.
74. J. Logan, *China Old and New* (Hong Kong: South China Morning Post 1982), p. 55.
75. S. Cochran, *Big Business in China: Sino-Foreign Rivalry in the Cigarette Industry, 1890–1930* (Cambridge, MA: Harvard University Press 1980).
76. I. Cheng, V.L. Ernster and H. Guanqing, 'Tobacco Smoking among 847 Residents of East Beijing, People's Republic of China' in *Asia-Pacific Journal of Public Health*, 4, 1990, pp. 156–63.
77. See D. Bello, *Opium and the Limits of Empire: Drug Prohibition in the Chinese Interior, 1729–1850* (Cambridge, MA: Harvard University Press 2005).

3
Developing Habits: Opium and Tobacco in the Indonesian Archipelago, c. 1619–c. 1794

George Bryan Souza

Introduction

This chapter focuses on the role played by the Dutch East India Company (VOC) in the maritime trade in opium and tobacco in South and Southeast Asia in the seventeenth and eighteenth centuries.[1] Batavia, the colonial port city market on Java, became a center for commerce in both products in this period. This commerce is traced using the Dutch Company's impressive archives to form a picture of how and when trading activities in opium and tobacco developed and the extent to which the fate of each product was intertwined with that of the other.[2]

A focus on Batavia in this period makes sense as the VOC was a significant early modern commercial organization with global reach and connections. Recognizing the imperative for the Company to possess and develop a port city independent of Asian interference, J.P. Coen,[3] the iron-willed Dutch colonial administrator, ordered the seizure of an appropriate site on the Java coast in 1619. Known as Jakarta, the Dutch re-named the site Batavia. Batavia was the archetypical colonial port city in Asia in that it served as a gateway 'through which European power and influence, economic dominance and technological modernization (to which could be added disease and health care) flowed from overseas to the furthest corners of the continent and as primate cities controlling their hinterlands and acting as vital lynchpins in the development of the world economic system'.[4] The Company instigated policies that attracted foreign merchants, European and Asian, to trade at the port which became an important nexus for regional, inter-regional and global maritime trade in the seventeenth and eighteenth centuries.[5] After 1684 the Company may be characterized as more pragmatic than bellicose in its commercial inter-relationships with other states and Asian and European merchants.[6] By 1792 the Company's commercial activities, as distinct from their political and colonial policies in the Indonesian archipelago, had been surpassed by European competitors.[7] The history of the acceptance and use of tobacco and opium in the region over this period had an enduring significance. The Dutch would use and employ that legacy in the nineteenth century, by establishing the controlled sale or farming of opium to provide revenue to finance the expansion of their colonial administration throughout the Indonesian Archipelago.

Batavia and opium purchasing

After early experiments with exporting small quantities of Malwa opium from Surat to Batavia, the Company decided to concentrate its activities related to this commodity in Bengal. The VOC initiated its long commercial relationship with opium from that region in the 1650s when the Company began to export Bihar opium from Bengal to Malabar and Batavia.[8] VOC exports to the Malabar coast were used to acquire pepper and those to Batavia were for sale, profit and re-distribution in the Indonesian archipelago, the Malay world, the rest of mainland South-east Asia[9], and China.[10] Malabar was the primary market destination for VOC opium exports from Bengal until the 1670s. After that date the Company and its competitors continued to export Bengal opium to Malabar, but VOC exports from Bengal overwhelmingly favored Batavia.[11]

The VOC established their primary factory at Hugli and an additional unit at Patna as part of their overall commercial strategy in Bengal in order to obtain opium grown in Bihar. The company estimated in 1688 that 'the annual output of opium in Bihar in a normal year at 8700 *maunds* (595,950 *ponden* or 4350 chests)'.[12] One *maund* was 681/2 *ponden* and one Dutch *pond* was equal to 1.09 lbs or 0.4 kilos. This same report identified the producers, the extent of their holdings, and their socio-economic and political positions. It estimated that at this time 11–13 per cent of Bihar opium production was consumed locally in Bengal, and another 34–46 per cent was sold at Agra and Allahabad, while the remaining 41–55 per cent was exported to other markets in India and abroad. The VOC's participation '. . . around this time was approximately 1000 maunds (500 chests), accounting for about 11.5 per cent of the total output'.[13]

Long before the Battle of Plassey in 1757 and the implementation in 1773 of the British East India Company's (EIC) monopsonistic policies toward the procurement of opium by other traders,[14] the VOC encountered and managed a variety of political and economic difficulties in their activities in Bengal. One of their foremost problems was competition in the opium trade. This came from Asian sources,[15] but also from rival European companies and private merchants that included the Portuguese,[16] the Danes,[17] and more significantly, the English and the French.[18] Since the Company could not establish a monopoly in Bengal over the supply of opium, it resorted to proclaiming and reserving the sale of opium throughout the Indonesian Archipelago and in parts of the Malay world as an exclusive right for the VOC. While the VOC made the claim of monopoly, this policy could only be implemented via negotiation with the different indigenous state systems in the region.[19] Their adherence to such agreements was not strictly enforced and was contested by Asian and European merchants, especially the British who were not party to any such agreements.[20] The burden of enforcement of these agreements and other ordinances or *plakkaaten*[21] fell to VOC naval forces and administrators who were obliged to apprehend violators[22] and then prosecute them.[23]

Over this entire period, the opium that the Company ordered, purchased and exported from Bengal was almost completely sourced from Patna in Bihar, although there were some purchases and deliveries from Balasore between 1749 and 1754. While there were differing gradations of opium, the Dutch reports do not distinguish or differentiate between them in their purchase or sales records. In the absence of written complaints of adulteration with sand, it appears that the quality of opium from Patna was accepted. Patna opium was sold and purchased by weight in lots of one *maund*. The rupee, the standard local silver coin, was used for payment, which permits its conversion to Dutch currency. The opium that was exported was presented in a standardized format, that is by chest. Each chest held two *maund* or 137 *ponden* of opium. Its weight including the contents was set at 145 *ponden*. Purchase and sales prices and quantities of opium were reported in chests, a practice that is followed throughout this chapter.

The Company determined the quantity of opium that it wished to be ordered in Batavia. This decision was made at the highest levels of the Company's administration and involved the Governor-General and Council, all experienced merchants, who were kept informed of prices and market conditions. Consequently the orders sent to Bengal may be interpreted as indications of the demand for opium at Batavia and throughout the consuming markets in the Indonesian Archipelago. This was not a unique practice. Other Companies, the Directors of the EIC for example, followed a similar practice in regard to commodities in their instructions to their supercargoes. There was, however, some flexibility for local Company merchants in the execution of the orders, depending upon the price and the capital that was available as well as the logistical issues associated with the physical delivery of the commodity.

The mode of procurement of Bengal opium by the VOC appears to have incorporated the services of one or more *dalal* (broker) who was 'a native employee with an intimate knowledge of both the local market and the merchants'.[24] These employees would invite reputable and reliable indigenous merchants to meetings in Hugli and Patna with the Company's Dutch merchants to negotiate supply contracts or spot sales of opium. Upon delivery to the Patna factory the Company would arrange fluvial transport from Patna to Hugli via indigenous shipping, and the opium was delivered for reception and subsequent loading and export by the Hugli factory on Company ships sailing from Bengal to Batavia. If necessary and if available, the Hugli factory could participate in this process by purchasing Patna opium at Hugli, usually at a higher price, to increase the quantity exported and attempt to fulfill orders. Generally, the annual exports of Bengal opium were delivered to Batavia from November to March.

Dutch records indicate that the Company's representatives encountered difficulties in acquiring the quantities of opium that Batavia ordered for the developing markets and consumer demand that they identified in the Indonesian Archipelago. Over the 112 years from 1659 to 1771, there are 86 years that reported both orders for and exports of Bengal opium. The VOC's merchants at Patna and Hugli failed more often than not in those years to obtain the quantities of opium that were sought for the market at Batavia. This evidence suggests that they were able to equal or exceed the amount of opium ordered only around a quarter of the time. The inability of the Company to get all of the opium that it wanted did not impede the VOC from benefiting and profiting from this trade. It simply made the opium business more complicated, costly, and competitive.

The VOC exported 67,831 chests of opium from Bengal to Batavia over this same 112-year period. Figure 3.1 demonstrates the rise and fall of the quantities of Bengal opium exported by the VOC to Batavia from 1659 to 1771. The average number of chests exported by the Company rose from nine chests in 1659 to 28 chests in 1666, for an annual average of 14 chests for those years. In the following decade from 1667 to 1676,

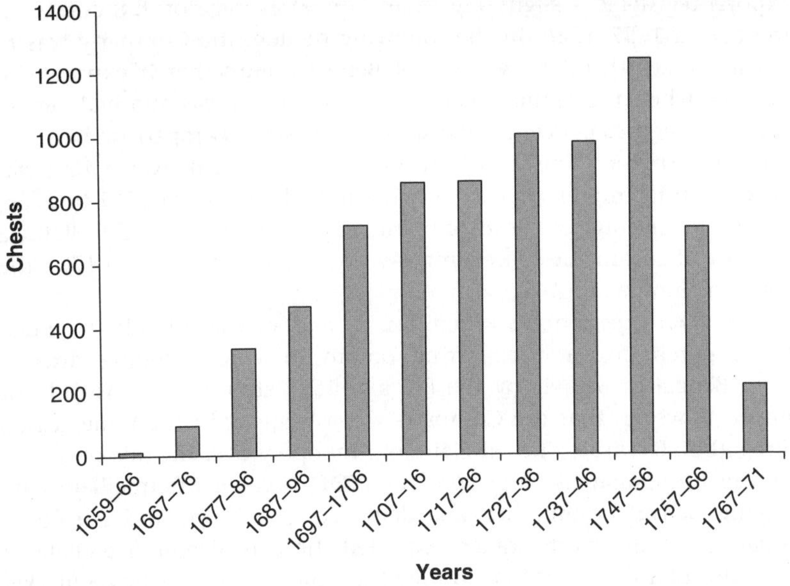

Figure 3.1 The Rise and fall of VOC exports of Bengal Opium to Batavia, 1659–1771 (average number of chests/year).

opium exports from Bengal almost quadrupled to an annual average of 92. Over the following decade from 1677 to 1686, exports of Bengal opium to Batavia more than tripled, to an annual average of 334 chests. Growth in Bengal opium exports to Batavia slowed but still grew over the decade from 1687 to 1696. Compared with the previous decade, they grew over 30 per cent to an annual average of 467 chests. The rise in Bengal opium exports continued and nearly doubled the number of chests in the decade from 1697 to 1706 in comparison with the previous decade, when an annual average of 718 chests was registered.

Over the next two decades, from 1707 to 1716 and from 1717 to 1726, the meteoric rise of Bengal exports to Batavia slowed, growing approximately 15 per cent in the decade from 1707 to 1716 compared with the decade of 1697–1706 and stabilizing at an annual average of 853 and 859 chests, respectively, for both the decades. Expansion in exports to Batavia still continued over the next decade, from 1727 to 1736, the annual average of 1005 chests was broached, nearing an inter-decennial growth rate of 20 per cent compared with the previous decade. The following decade was a period of consolidation and the Company's exports registered a slight slip to an annual average of 978 chests for the period 1737–1746. In the following decade, the Company was to reach the zenith of the volume of Bengal opium that it exported to Batavia, when the annual average of 1241 chests was attained; again, slightly over a 20 per cent expansion in volume in comparison with the previous 10 years. The fall of the VOC's exports of Bengal opium was faster than its impressive rise; over the next decade from 1757 to 1766, the annual average of chests of opium exported by the VOC to Batavia fell to 710 and in the succeeding 4-year period from 1767 to 1771 the number slumped to 216.

The VOC experienced difficulties in, and was eventually excluded from, purchasing and exporting opium in VOC shipping directly from Bengal to Batavia by the EIC's policies after 1757. This did not mean, however, that the Company's involvement in the trade ceased altogether. By virtue of their claim of monopoly over the sale of opium in the Indonesian Archipelago, the VOC became intermediaries by buying Bengal opium that was delivered, purchased and re-sold at Batavia. British private traders exported the Bengal opium to Batavia over this period, which transformed the structure of the Batavia market and prolonged the incorporation of stocks of opium from Bengal by the VOC. Although the Dutch Company's gross profit margins from this trade suffered, the arbitrage was favorable, which will be discussed when prices are mentioned below. From 1769 to 1789 the Company purchased 4766 chests of Bengal opium from British private traders at Batavia. There is no particularly striking trend in the quantities that the VOC was able to acquire by embracing this practice. There were interruptions in deliveries of Bengal opium to Batavia by British private traders that were registered in the years 1770–1771 and 1775–1781; the latter was caused by the support of The Netherlands for the rebel position in the American Revolutionary War. The VOC was able to obtain an annual average of 156 chests from 1769 to 1774 and to increase this number substantially to an annual average of 541 chests from 1782 to 1787. The annual average for the following 2 years from 1788 to 1789 fell to 294 chests.

Batavia and opium distribution

The opium exported from Bengal on VOC shipping was received at Batavia and warehoused. The Company endeavored to move all of the stock of opium that was received on an annual basis. When market or other conditions made this impossible, stocks were carried over into the following year. From 1659 to the late 1680s the Company mixed several practices in its channels of distribution of Bengal opium on Java and throughout the Indonesian Archipelago. The VOC on occasion would load opium either in Bengal or at Batavia and incorporate it into direct trading ventures to ports on Java's north coast.[25] They would also sell opium by public auction at Batavia. By the late 1680s the Company desisted from the practice of direct VOC voyages and sales in preference to regular sales at public auction at Batavia to indigenous merchants (Indonesians, Malays, or overseas Chinese) and foreign traders (primarily Portuguese from Macao, Chinese, Siamese, Armenians, and Spanish). The use of the public auction at Batavia was to remain the Company's exclusive channel of distribution for Bengal opium until 1745. In that year the VOC made a major shift in its commercial policy approving the formation of the *Amphioen Sociëteit* or Opium Society at Batavia.[26] A company within the Company, the Opium Society took over the responsibility of developing and financing the sales of opium to indigenous merchants from the VOC in exchange for a guaranteed price on delivered volumes. The *Amfioen Sociëteit* continued its activities until 1794, the end of the period under study. While the establishment of the Society offered the advantages of diminishing price volatility for the Company and the transfer of a set of commercial and administrative functions, additional research in the accounts of both the Company and the Society is required to determine whether this was a beneficial decision for the VOC.

The quantities of opium that the Company sold and the prices that it realized from these sales entered the VOC's accounts in its annual income statements for Batavia, which were entered from the 1st of September to the following 31st of August. This data includes the opium sold by the Company directly and via public auction from 1673 to 1745 and to the *Amfioen Sociëteit* from 1745 to 1789. It details the Company's gross profit margin calculations on the sales and summarizes the purchases and re-sales of opium from Bengal from British private traders from 1769 to 1789. Prior to the 1750s, all of the opium sold by the Company was from Bengal. According to Baud in 1753 and 1778, small quantities (90 chests in both the years) of opium from the Levant via Amsterdam was also sold by the VOC to the *Amfioen Sociëteit* at Batavia.[27]

Based upon the data from 1673 to 1789, a trend analysis may be created for the quantity of opium sold by the VOC and the *Amfioen Sociëteit* to indigenous and foreign merchants at Batavia from 1688 to 1789. Figure 3.2 presents this analysis. Over the irregular number of

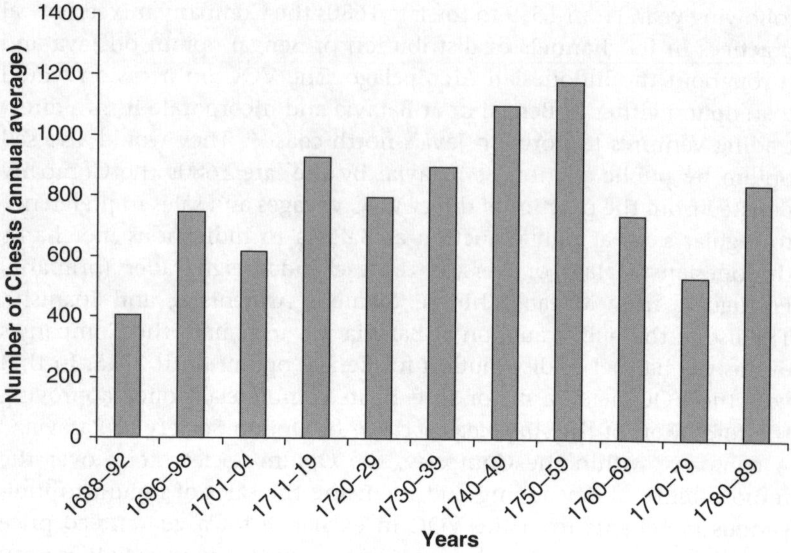

Figure 3.2 Opium sales at Batavia, 1688–1789.

reporting years, as may be observed, for the quantity of opium sold from 1688 to 1692, 1696 to 1698, 1701 to 1704, and 1711 to 1719, the VOC sold on an average annual basis, respectively, 403, 745, 617, and 928 chests. In the following decades from 1711 to 1789, the VOC sold on an average annual basis, respectively, 928, 795, 897, 897, 1100, 1180, 739, 537, and 845 chests. The Company reached the apogee of its sale of opium in the years after making the decision to shift the channel of distribution from public auction to the *Amfioen Sociëteit*. Subsequently, opium sales declined because of supply difficulties in Bengal due to the slow but effective implementation of EIC policies and the prevailing wartime conditions of the period from 1775 to 1781.

For the VOC, as well as the indigenous growers and merchants involved in the production of opium in Bihar and Bengal, the prices that this commodity commanded were intrinsic to the profits that could be realized in its commercialization at Batavia. The same was true for the indigenous and foreign merchants that purchased the commodity from the VOC at Batavia and re-distributed and sold it for profit in the consuming port cities of Java and others throughout the Indonesian Archipelago, the Malay World, and China. The profit motive was present for all participants along the supply chain. How profitable opium actually was for those involved in Bihar and Bengal is difficult to determine. The Company's records do provide a set of data for the prices that it paid for opium in Bihar and Bengal, as well as what they paid British private traders for Bengal opium in Batavia and the price that the Company and the *Amfioen Sociëteit* re-sold opium, throughout the long eighteenth century. The same records also provide sporadic data for the prices obtained for Bengal opium by indigenous merchants at a few of the markets where it was sold on Java.

Based on the data that is presently available, for the period from 1672 to 1717, the VOC paid from 60 to 170 *rsd.*/chest of opium in Bengal and sold this opium at Batavia for 200–572 *rsd.*/chest. Over the same time span the Company reported that the indigenous merchants who bought and re-distributed this opium at different port-cities on Java were selling the same opium at 30–70 *rsd.*/chest more than what they had paid the Company at Batavia. Over the period from 1719 to 1754, the average annual price that the Company paid to purchase opium in Bengal was 178 *rsd.*/chest. At the same time the average annual price for which the VOC sold the same Bengal opium at Batavia was 394 *rsd.*/chest.[28] The same report indicates that the Company calculated its annual gross profit margins in opium from the exceptionally and uniquely low margin of 18 per cent in 1725 to the exceptionally high margin of 271 per cent in 1743. According to Baud and the primary records the Company reported that it earned 48,540,377 *florins* in profit from the sale of opium from 1677 to 1789.[29] This figure may be converted to 16,180,125 *rsd.*, which is equal to 19,416,150 Spanish *pesos* and 17,651,046 Chinese *taels*. It is not possible, at this moment, to serially document the prices or margins that indigenous merchants were obtaining for the Bengal opium that they were selling throughout the Indonesian Archipelago, the Malay World, and China. Research is ongoing on this issue. The same is true for the accounts of the *Amfioen Sociëteit*. Preliminary research in those accounts suggest that the prices and the margins that the *Amfioen Sociëteit* enjoyed were substantial. This would suggest that the *Sociëteit* pressured prices upwards and negatively impacted on the margins of indigeneous intermediaries and on the final price to consumers. During the last period, from 1769 to 1789, in which the Company bought Bengal opium from British private traders at Batavia and then re-sold the same, the annual arbitrage profits for the Company ranged from a miniscule 13 *rsd.* to 609 *rsd.*/chest in favor of the VOC. In 2 years, 1772 and 1774, the Company lost 80 and 100 *rsd.*/chest on the Bengal opium that they had purchased from British private traders.[30]

Tobacco and opium in the Indonesian Archipelago

Throughout the sixteenth and seventeenth centuries, New World tobacco was introduced via contacts and exchanges with the Portuguese and Spanish in Asia.[31] The dissemination and acculturalization of tobacco production and consumption was widespread and profound. Previous social customs and practices eased this, as the practice of betel-chewing served as the cultural bridge to tobacco smoking throughout the Indonesian Archipelago and the Malay World.[32] Tobacco 'gradually came to fill a similar [to betel-chewing] combination of relaxant, social, and medicinal roles... [it] appears at this stage to have been regarded as a stronger and more expensive alternative to betel'.[33] Its popularity grew in the seventeenth century through the smoking of cigars or *bungkus* (bundles), which were homemade from 'shredded tobacco, often mixed with other aromatics or flavorings'.[34] By the end of the seventeenth century, the popularity of *bungkus* was such that Europeans reported that men, women, and children in the region were addicted to the nicotine habit.[35] In addition to the form of consuming tobacco by smoking cigars, the smokeless practice of chewing was known, but it did not increase in popularity until the eighteenth century.

Smoking tobacco in pipes, which emulated contemporary European practice, competed with the popularity of cigar smoking in the seventeenth and eighteenth centuries in the Indonesian Archipelago and the Malay World. As with cigar smoking the acceptance of pipe smoking also derived from it being seen as an alternative to betel-chewing. Although indigenous consumers emulated European practice, there was a significant and an extremely important difference in its acculturalization. Pipe smokers throughout the Indonesian Archipelago and the Malay World added opium to the tobacco that they were inhaling. The mixture of tobacco and opium was known as *madat*. The smoking of *madat* was first reported by the Dutch on Ternate in 1610[36] and the smoking of cigars or *bungkus* in 1658.[37] The first *plakaaten* prohibiting and heavily fining both the vendor and the consumer of *madat* in the areas of jurisdiction of the Company was promulgated on 22 December 1671. This practice was neither extinguished nor diminished by this edict, which the Company reinforced on 28 December 1729, when it doubled the already heavy fines.[38]

Concurrent with the emergence of this new mode of consuming opium and tobacco together was an expansion in the production of tobacco throughout the Indonesian Archipelago, the Malay World, the Philippines, and China. Chinese junks and the Portuguese from Macao initiated the export of Chinese tobacco to Batavia, where it was sold to the VOC. The foreign traders (Chinese, Siamese, Portuguese, Spanish, and Armenians) active in the trade to and from Manila included Philippine tobacco production in their cargoes. It is possible to quantify from Dutch records the amounts of Chinese and Manila tobaccos that were involved in this trade. But, for the sake of brevity, they are not provided in this chapter. By the late seventeenth and early eighteenth century, tobacco produced on Java and elsewhere in the Indonesian Archipelago had entered the circuits of commercialization. But Chinese tobacco dominated the transactions and maritime trade in this commodity. This product was preferred by, and destined for, the burgeoning overseas Chinese market and was also sold to local communities throughout the region.

At Batavia, where exports of Chinese tobacco and Bengal opium arrived and were sold by public auction, VOC authorities prepared reports on the arrival and departures of indigenous and foreign vessels that traded in these and other commodities. The number of vessels and the port cities from which they arrived, and to which they were headed, was also recorded. The volumes of these exports were not quantified in this report but their value was provided in *rijksdaalers*. These reports were incorporated at the end of every month into the Batavia *dagregisteer*. They permit the reconstruction of the channels of re-distribution by indigenous and foreign merchants shipping and give a sense of the markets for which that Chinese tobacco and Bengal opium were destined.

A trend analysis may be generated for the values and composition of goods that were purchased at Batavia and redistributed by indigenous and foreign merchants in eleven 5-year periods from 1670 to 1720. Table 3.1 presents this data. Research is ongoing to complete this analysis for the entire period. The total values include goods and commodities that were exported by foreign merchants to mainland Southeast Asia, the Philippines, and China. They distort the analysis slightly because they include non-regional destinations. There is less interpretative difficulty with textiles, opium, and tobacco, since they were overwhelmingly and, at this time almost exclusively, on board vessels that were destined for the Indonesian Archipelago and the Malay World. The textiles were entirely Indian cottons from diverse subregional sources of supply, since Chinese raw silk and piecegoods and Dutch textiles were placed in the 'others' category. The opium was from Bengal and the tobacco values are Chinese in origin. When the values of non-regional exports are suppressed, the percentage participation of

Table 3.1 Small Vessels Report: Annual departures from Batavia, goods and values, 1670–1720.

Year	Numbers of Vessels	Cotton cloth (diverse)	Opium	Tobacco (Chinese)	Others	Total Value (*rsd.*)
1670	1,152	135,763	1,731	–	43,431	180,925
1675	1,152	57,653	15,882	20	74,736	148,291
1680	1,063	87,396	26,386	371	77,849	192,002
1685	1,522	69,093	31,541	1,397	77,325	179,356
1690	1,482	53,148	135,754	9,275	84,218	282,395
1695	1,336	40,800	132,781	7,317	156,892	337,790
1700	1,009	19,072	40,440	2,818	190,985	253,315
1705	779	13,149	61,165	2,010	128,735	205,059
1710	731	21,215	31,266	1,528	37,240	91,249
1715	751	16,237	65,388	3,093	140,529	225,247
1720	801	72,682	110,196	10,541	149,299	342,718
Total:	11,778	586,208	652,530	38,370	1,161,239	2,438,347

Sources: ARSIP, DRB, Vessels Report found at the end of each month for the calendar year.

opium to total value and other exports to the Indonesian Archipelago is commensurately higher.

Some of the more notable trends are as follows. The value of Bengal opium exports eclipsed those of Indian cotton cloth by 1690. The value of opium exports retained its supremacy for the rest of this period and probably for the entire period. A quarter of the revenue of all goods and commodities that the VOC sold to indigenous and foreign merchants was from Bengal opium. That percentage is even higher, it is estimated at one half or more, when compared with the value of the goods and commodities that the VOC sold that were destined exclusively for markets in the Indonesian Archipelago. In 1695 for example, 119,963 *rsd.* worth of pepper, cloves, nutmeg, and sandalwood was purchased and exported to Macao, other South China ports, and Vietnam. By suppressing those sales and exports, the total value of commodities sold and redistributed in the Indonesian Archipelago would have been 217,827 rsd., the sales value of opium was 132,781 rsd., almost 60 per cent of the modified value and textiles were 40,800 rsd., around 20 % of the modified total. Finally, while the value of purchases and exports of Chinese tobacco was small, there are indications of a seemingly symbiotic market relationship between Chinese tobacco and opium. Chinese tobacco was almost entirely exported to markets where opium was also being exported. It was only on rare occasion and in small value that Chinese tobacco was exported to a market that was not receiving opium.

Using this same data, it is also possible to create a trend analysis by geographical end-market for the goods and commodities that were purchased at Batavia and exported and redistributed by indigenous and foreign merchants in 11 years, from 1670 to 1720. Once again, further research will take this analysis to the end of the entire period. The following are the observable trends that are suggested by this data. The total value of opium that was redistributed over these 11 years from 1670 to 1720 was 652,650 *rsd*. The indigenous merchants in their vessels redistributed the vast majority, around 90 per cent, of this opium on the island of Java. Of this 90 per cent, most of the Bengal opium was destined for the Java North Coast. When the VOC records provide greater detail on specific port city destinations on Java, opium was redistributed in descending values to: Semarang, Rimbang, Cirebon, Gresik, and Surabaya. Of the other 10 per cent, half of it was delivered to undecipherable destinations because of damage to the documentation and the other half or only 5 per cent of the total was redistributed in descending values to: Borneo (Banjarmasin), Sumatra (Palembang), Sulawesi (Macassar), Lesser Sunda Islands (Timor), Moluccas (Ambon), China (Macao), Mainland Southeast Asia (Thailand), and the Malay World (Jambi and Johor). Exports to China via Macao were only registered once in 1720. While the concentration on Java may have been expected from secondary reading, it is remarkable how little opium in value or quantity was destined by this channel of redistribution to the Malay World, which may indicate how competitive the British and other indigenous merchants were in their efforts in those markets.

Conclusion

This chapter has shown how the Dutch VOC played an important role in the supply of opium and tobacco to Southeast Asian markets in the seventeenth and eighteenth centuries. As regards opium, it has presented a sequential supply chain that flows from the production of opium in Bihar to its final consumption throughout the Indonesian Archipelago, the Malay World, mainland Southeast Asia, and China via VOC operations in India and at Batavia. The VOC often faced difficulties in procuring opium, and faced competition in selling it, even in areas over which it claimed a monopoly. Despite this, the sale of Bengal opium produced significant income and profits for the VOC. Yet it must be remembered that the VOC was only one link in a chain that included a range of local agents. Indian producers in Bihar used the Dutch to secure markets outside of South Asia for their products. The fundamental role of indigenous merchants and shipping in the re-distribution of important values of Bengal opium and Chinese tobacco to the same consuming markets on Java was also highlighted.

Demand for opium grew in this period. The VOC characterized consumer demand for opium at Semarang on Java as good in the early 1680s because 'it appears that the Javanese cannot live without it'.[39] The VOC was convinced that the strong demand for opium on Java at Mataram in the early 1680s was caused by wartime conditions, which suggests that *madat* may have been employed as payment for, or in the provision of victuals to, soldiers. This speculation has yet to be confirmed.[40] However, it is clear that demand for opium waxed but rarely waned over the long eighteenth century and that it was established in times of peace as well as war.

The chapter has also identified an interesting relationship between tobacco and opium sales. While research in this direction is in its early stages, it is certainly the case that between 1660 and 1720, markets for opium also became markets for tobacco and that over this period demand for opium at those markets increased as demand for tobacco increased. There has been little attention paid to the cultural practices and habits on Java, in the Indonesian Archipelago and elsewhere in the Malay world of Southeast Asia and in China to explain this rise in consumption of opium over the later half of the seventeenth and the first half of the eighteenth centuries.[41] It may well be that this link with tobacco holds the key to an explanation as it has been shown elsewhere that the introduction of tobacco changed consumer habits relating to opium.[42] While it had been most commonly ingested, the practice of smoking that came with tobacco introduced users to the idea of burning opium; at first as an addition to the tobacco mix but later as a powerful pipe in its own right. The practice of ingesting opium came with its own natural checks on demand as the body could only take so much before sleep or nausea halted consumption. Smoking, however, relaxed such checks and as such is likely to have created a greater demand for the product. It was certainly the case that the Dutch had noticed that opium was being mixed with tobacco as early as 1610 and that edicts against the practice in VOC territories later in the seventeenth and early in the eighteenth centuries confirm that the habit endured. Increasing sales of opium by the VOC throughout the period covered in this chapter may, at least in part, therefore simply reflect changes in consumer habits in Southeast Asia.

Notes

1. For recent scholarship on opium and empires, see C. Trocki, *Opium, Empire and the Global Political Economy: A Study of the Asian Opium Trade 1750–1950* (London: Routledge 1999); J.F. Richards, 'Opium and the British Indian Empire: The Royal Commission of 1895' in *Modern Asian Studies*, 36, 2, 2002, pp. 375–420; A. Farooqui, *Smuggling as Subversion: Colonialism, Indian Merchants and the Politics of Opium* (Oxford: Lexington 2005).
2. The VOC records are found in the General State Archives (the *Algemeen Rijksarchief (ARA)*) in the *Koloniale Archieven Oost-Indie: Archieven van de Vereenigde Oost-Indische Compagnie* (VOC) collection and papers in The Hague; for the printed Dutch records that were used, see W.P. Coolhaas *et al.* (eds), *Generale Missiven van Gouverneurs-General en Raden aan Heren XVII der Verenigde Oostindische Compagnie*, 11 vols (The Hague 1960–2001) abbreviated and cited in the text as *GM*; J.A. van der Chijs *et al.* (eds), *Daghregisteer gehouden int Casteel Batavia vant passerende daer ter plaetse als over geheel Nederlandts India*, 31 vols (Batavia and The Hague 1888–1931) abbreviated and cited in the text as *DRB*; the post-1683 *Daghregisteer* of Batavia documentation was consulted in microfiche at the General State Archives. The originals are found in the Indonesian National Archives (ARSIP) and are cited as ARSIP DRB.
3. See H.T. Colenbrander and W.P. Coolhaas, *Jan Pietersz. Coen, Bescheiden omtrent zijn bedriif in Indie*, 7 vols (The Hague 1919–1953).
4. See F. Broeze, 'Introduction: Brides of the sea' in F. Broeze (ed.), *Brides of the Sea: Port Cities of Asia from the Sixteenth to the Twentieth Centuries* (Honolulu: University of Hawaii 1995), p. 4.
5. See K. Glamann, *Dutch Asiatic Trade, 1620–1740* (The Hague 1958); J.R. Bruijn *et al.*, *Dutch-Asiatic Shipping in the 17th and 18th Centuries*, volume I (The Hague 1979–1987), pp. 119–42, 195–209, and 223–45.
6. See B.B. Kling and M.N. Pearson (eds), *The Age of Partnership: Europeans in Asia Before Dominion* (Honolulu: University of Hawaii 1979).
7. See H. Furber, *Rival Empires of Trade in the Orient 1600–1800* (Minneapolis: University of Miinesota Press 1976).
8. For opium commercialization in Bengal and Indonesia, see O. Prakash, *Dutch East India Company and the Economy of Bengal, 1630–1720* (Princeton: Princeton University Press 1985); J.C. Baud, 'Proeve van eene Geschiedenis van het Handel en het Verbruik van Opium in Nerderlandsch Indie' in *Bijdragen tot de Taal, Land, en Volkenkunde van Nederlandsch Indie*, 1, 1853, pp. 79–220; F. de Haan, *Priangan: De Preanger Regentschappen onder het Nederlandsch Bestuur tot 1811* (Batavia 1910–1912).
9. For general comments on opium use in the Indonesian Archipelago, the Malay world and mainland South-east Asia, see M.C. Ricklefs, *A History of Modern Indonesia Since c.1200*, third edition (Stanford: Stanford University Press 2001), pp. 87, 99, 101, 104–6, 112, 118, 159, 162–3, and 173–6; Anthony Reid (ed.), *Southeast Asia in the Early Modern Era: Trade, Power, and Belief* (Ithaca: Cornell University Press 1993), pp. 112, 115, and 134; L.Y. Andaya and B.W. Andaya, *A History of Malaysia*, second edition (Honolulu: University of Hawaii Press 2001), pp. 92 and 103–4. Recent work on the VOC and contacts with indigenous state systems in the Indonesian

Archipelago, which incorporates discussion of opium commercialization, includes R. Vos (translated by B. Jackson), *Gentle Janus, Merchant Prince: The VOC and the Tightrope of Diplomacy in the Malay World, 1740–1800* (Leiden: KITLV Press 1993). For the VOC's tade relations with Palembang, especially in pepper and tin, see D.N. Lewis, *Jan Compagnie in the Straits of Malacca, 1641–1795* (Athens: Ohio University Press 1995). For Melaka, see L. Nagtegaal (translated by B. Jackson), *Riding the Dutch Tiger: The Dutch East Indies Company and the Northeast Coast of Java, 1680–1743* (Leiden: KITLV Press 1996). For the VOC's relations and trade with the northeast coast of Java, see Gerrit J. Knaap, *Shallow Waters, Rising Tide: Shipping and Trade in Java around 1775* (Leiden: KITLV Press 1996). For indigenous Javan port cities that were involved in intermediate and coastal maritime shipping and marketing activities, see Gerrit J. Knaap and Heather Sutherland, *Monsoon Traders: Ships, Skippers, and Commodities in Eighteenth-Century Makassar* (Leiden: KITLV Press 2004).

10. For trends in the trade of Indian opium to China, see H.B. Morse, *The Chronicles of the East India Company Trading to China, 1635–1834* (Oxford 1926–1929, reprint Taiwan 1975); F. Wakeman, 'The Canton Trade and the Opium War' in J.K. Fairbank (ed.), *Cambridge History of China*, volume 10 (Cambridge: Cambridge University Press 1978), pp. 163–212; K. Pommeranz and S. Topik, *The World That Trade Created: Society, Culture, and the World Economy, 1400-the Present* (London: Sharpe 1999), p. 103. For the use of opium and tobacco in China, see M. Heijdra, 'The Socio-Economic Development of Rural China During the Ming' in D. Twitchett and F.W. Mote (eds), *The Cambridge History of China*, volume 8 (Cambridge, 1998), pp. 417–578; J.D. Spence, 'Opium Smoking in Ch'ing China' in F. Wakeman Jr. and C. Grant (eds), *Conflict and Control in Late Imperial China* (Berkeley: University of California Press 1975), pp. 143–73; R.K. Newman, 'Opium Smoking in Late Imperial China: A Reconsideration' in *Modern Asian Studies*, 24, 4, 1995, pp. 765–94; Y. Zheng, 'The Social Life of Opium in China, 1483–1999' in *Modern Asian Studies*, 37, 1, 2003, pp. 1–39.
11. O. Prakash, *Dutch East India Company*, p. 57.
12. Ibid.
13. Ibid., p. 58.
14. See O. Prakash, 'Trade and Politics in Eighteenth Century Bengal' in L. Blussé and F. Gaastra (eds), *On the Eighteenth Century as a Category of Asian History: Van Leur in Retrospect* (Aldershot: Ashgate 1998), pp. 237–60.
15. See *GM*, III, p. 547 for the incorporation of Bengal opium in Mughal trading activities with neighboring Arracan in the 1660s; for details of the activities of indigenous and European traders involving purchases of Bengal and sales at Bantam in the late 1670s and early 1680s prior to the Company's occupation of that port-city, see *GM*, IV, pp. 18, 389, and 402; J.M.J. de Jonge, *De Opkomst van het Nederlandsch Gezag in Oost-Indië*, volume VII (The Hague/Amsterdam 1862–1888), pp. 9–10. For VOC complaints of Malay involvement in the trafficking of Bengal opium to Andragieri, Jambi, Palembang, Borneo, and the ports of the Java north coast in the 1700s, see *GM*, VI, p. 431.
16. For an example of the Portuguese involvement in exporting Bengal opium to the Malabar Coast in the 1690s, see G.B. Souza, 'Portuguese Colonial

Administrators and Inter-Asian Maritime Trade: Manuel de Souza de Menezes and the Fateh Moula Affair' in *Portuguese Studies Review*, 12, 2, 2004.

17. See *GM*, V, pp. 758–61.
18. While sufficient evidence has not been located to establish a reliable trend analysis, in specific years English and French purchases and exports of opium from Bengal were superior or significant in comparison to the VOC's purchases and exports. For example, the VOC reported that English purchases and exports in 1711 were 850 chests in a year that the VOC exported 800 chests and that French purchases and exports in 1714 were 4–500 chests in a year that the VOC exported 1165 chests, see: *GM*, VI, p. 719; *GM*, VII, pp. 105–6.
19. See J.E. Heeres and F.W. Stapel (eds), *Corpus-Diplomaticum Neerlando-Indicum* (The Hague 1907–1953). For the agreements granting the VOC the exclusive right to import Indian textiles and opium into, for example, Mataram (1677), Palembang (1678), and Cheribon (1681), see *Corpus*, III, pp. 74–79, 140–2, 233–40, and 267–70. The VOC celebrated a similar treaty with Jambi in 1684, see *GM*, IV, p. 724.
20. Despite the treaty, the VOC reported indigenous shipping from Melakka north of Palembang in 1684 laden with textiles and opium, see *GM*, IV, p. 719.
21. See J.A. van der Chijs (ed.), *Nederlandsch-Indisch plakaatboek, 1602–1811* (Batavia: Landsdrukkerij, 1885–1900), III, pp. 229 and 535–6; IV, pp. 317–9, and 423–4; V, pp. 103 and 323.
22. Dutch naval forces stopped a French ship at Phuket (Udjung Salang) in 1692 and confiscated its cargo of opium, see *GM*, V, p. 591. The ship owned and captained by a Scottish merchant, Alexander Hamilton, was stopped and searched in 1712, and 18 chests of opium were confiscated by the VOC, see *GM*, VI, p. 858.
23. The Hamilton case was reviewed by the *Raad van Justitie* at Batavia in 1713 and the actions of the VOC authorities were upheld, see *GM*, VI, p. 912.
24. See O. Prakash, *Dutch East India Company*, p. 102.
25. See, for example, one of the Company's last trading ventures of this type to Jepara by the *Egmont* in 1685, *GM*, IV, pp. 773 and 812.
26. Relevant here are I.E. Mens, *De Amphioen Sociëteit (1745–1794). Middel tot 'redres' van de Compagnie of wellicht meer een middel to verijking van de Hoge Regering* (unpublished Ph.D. dissertation, Leiden University 1987) and S. van Galen, *Opium in de intra-Aziatische handel 1700–1760* (unpublished Ph.D. dissertation, Leiden University 1995). However, consultation of both has not yet been possible.
27. See J.C. Baud, 'Proeve van eene Geschiedenis', p. 151.
28. For the *pice* data used in this paragraph, see *GM*, III, pp. 853 and 906; IV, pp. 118, 163, 389, 413, 442, 663, 727, and 769; V, pp. 88, 274, 415 and 727; VI, pp. 165, 224, 262, 271, 335, 407, 515, 548, 719, and 772–3; VII, pp. 17–8, 71, 143, 271, 402, 465, and 592.
29. See J.C. Baud, 'Proeve van eene Geschiedenis', pp. 216–9.
30. See the Purchase and Sales Report at Batavia in the ARA, VOC, pp. 3250–822.
31. See B. Laufer, *Tobacco and its Use in Asia* (Chicago: Field Museum of Natural History 1924); T.O. Höllman, *Tabak in Südostasien: Ein ethnographisch-historischer Überlick* (Berlin: Dietrich Reimer Verlag 1988).

32. See A. Reid, 'From Betel-Chewing to Tobacco Smoking in Indonesia' in *Journal of Asian Studies*, 44, 3, 1985, pp. 529–49.
33. See A. Reid, *Southeast Asia in the Age of Commerce, 1450–1680*, volume I (New Haven and London: Yale University Press 1988), pp. 44–5.
34. Ibid., p. 44.
35. Ibid., p. 45.
36. J.C. Baud, 'Proeve van eene Geschiedenis', p. 88.
37. See A. Reid, *Southeast Asia in the Age of Commerce*, I, p. 44.
38. See J. Chijs, *Nederlandsch-Indisch plakaatboek*, II, p. xxx; IV, p. xxx. In 1671 the fines for first time offenders were 20 rsd. for sellers and 10 for buyers; for the second offense 40 rsd. and 20 for each; and on the third offense, the case was presented to colonial authorities for adjudication. In 1721, the 1671 fines were increased.
39. See *GM*, IV, p. 473.
40. See *GM*, IV, p. 673; J.M.J. de Jonge, *De Opkomst*, VIII, pp. v–vi.
41. See footnote 9.
42. See footnote 10.

4
Early British Encounters with the Indian Opium Eater

Richard Newman

Introduction

John Grose, who visited India in the 1750s, tells the story of an Indian nawab who invited the staff of a nearby English trading post to a day of relaxation and entertainment. During the afternoon a young writer, sauntering in the garden, picked the capsule off a poppy and, seeing that it had some beads of sap on its surface, he licked them off. Quite soon he felt lethargic and, on sitting down, fell into a coma. When told of this the nawab hurried across and anxiously enquired where the poppy had been picked. Grose reports that, on being told, the nawab 'apologised for having supposed that the nature of the poppies was in general too well known to have needed any warning... but that this particular sort... admitted of no human remedy... so that nothing could save him'. Indeed the young man died.[1]

This story should sound a note of caution, not just to callow young clerks, but to historians of drugs and empire. It is a reminder that, first, that the problem of controlling drugs has a history that goes back long before De Quincey and the Earl of Mar made it an issue in Britain, and the opium trade and the Opium War made it an issue in Asia. Moreover, territories that were eventually colonised by Europeans often had centuries of experience of psychoactive substances and a widespread understanding of their mental and physical effects; a knowledge that far surpassed that of their eventual imperial masters. This was certainly true of opium, India and the British East India Company. When the Company began to take territorial control of the subcontinent in 1765, it found itself the ruler of a society in which poppies were cultivated in several regions and the use of opium was subject only to the social controls which Indians had evolved for themselves.

By 1816 the Company had imposed a system of legal controls, which limited and taxed the supply of all intoxicating substances, opium included. This chapter traces the evolution of the system, examines the presuppositions which Company bureaucrats brought to the problem of opium regulation and considers how their policy changed in the light of experience.

East India Company encounters with opium before 1800

The first Europeans to describe the uses of opium in India were the Portuguese. They noted that it was widely eaten by rich and poor, in places as far apart as Gujarat, Bengal and the Malabar coast. Most people consumed small quantities and valued it because it took away their troubles and 'provoke[d] them to lewdness', but they were likely to increase the dose, become addicted and go about drowsy and confused.[2] The same points were made by later commentators from other countries. Linschoten, for example, stressed the aphrodisiac appeal of the drug while pointing out that men who 'eate much thereof, are in time altogether unable to company with a woman', a danger that led wealthy Indians to use it with caution.[3] Sexual pleasure apart, these early reports say more about the consequences of opium use than the motives for taking it.

From the early 1600s the East India Company began to understand why opium was so widely used. William Hawkins, on his arrival in Surat, called to see the local governor and was told he was not well, 'but I think rather drunk with affion or opion, being an aged man'.[4] This is interesting for both the information and the mitigation, for it shows recognition of the fact that opium might be used as a remedy for the aches and pains of old age. Hawkins moved on to Agra and sent his superiors a very precise account of the emperor Jahangir's daily drug-taking: six draughts of alcohol each evening and a pill of opium to help him sleep.[5] It was reported that Indian soldiers commonly took opium on the eve of battle to give them courage (in fact, the effect of the drug was to dispel their anxiety) and it was customary in the Rajput courts of western India to offer opium to visitors as a sign of peace and friendship.[6] The labouring classes of Surat used it as a stimulant and the town's long-distance messengers set out on lengthy journeys with a portion of it 'which they think fortifies them, and by this means will keep on running, and dozing as it were at the same time with their eyes open, and without feeling the fatigues of the way'.[7] Craftsmen used it to steady their hands and aid their concentration. In the 1770s, James Kerr of the Bengal Medical Service visited the poppy-growing areas of Bihar and wrote a paper on the production of opium, with a brief sketch of its consumption; its good and bad aspects were, he said, already well known from European books and the native uses, as stimulant, medicament and intoxicant, were very similar, if 'bolder'. He also reported that it had been used to dull the pangs of hunger during the famine of 1770.[8]

Alongside these factual statements there were often, though not invariably, moral and social judgements of a crudely 'orientalist' character, derived from travels through Turkey and Persia to the subcontinent. Englebert Kaempfer showed these biases with a wholly negative statement of opium's effects, followed by descriptions of hook-swinging, drugged temple dancers and addicts running amuck.[9] Sir Thomas Herbert, writing more about Persia than about India, contributed to the idea that oriental constitutions were somehow different from those in Europe and that large, repeated doses 'make that familiar, which would kill us; so that their medicine, is our poyson'.[10] This idea became established in English scientific writings, such as Mead's treatise on poisons, which stated that Turks and Persians commonly eat a drachm or two (20–40 grains) at a time, 'for the Effects of It in Them are no other than downright Drunkenness'.[11] Grose reported that opium users were aware of the damage they might do to themselves 'but this is a consideration that weighs little with the generality [of Indians], always more actuated by any present favourable objects, than by a providential regard for remote ones'.[12] Some writers conjured up a fantasy from the *Arabian Nights* to depict the ideas in the minds of the opium-eaters as they slid into sleep.[13]

There were, nevertheless, a number of studies published in the second half of the eighteenth century which gave a more considered view of the sociology and pharmacology of opium. Careful accounts of Muslim societies, such as Russell's much-quoted *History of Aleppo*, made it clear that opium was not taken as often as was commonly supposed.[14] Dallaway agreed, and made the point that the addict was considered 'with as much pity and disgust' as the inveterate drunkard was in the West.[15] Edward Smyth, in a paper that would certainly have been read by the medical profession, questioned many of the assumptions about addiction by describing the day he spent with a heavy opium-eater in Smyrna. The man took a morning dose equal to about 30 grains, which gave his face a new air of life and brightness; he said he felt refreshed, went off to his coffee house, and half an hour later was cutting wood for his fire. He came in for a second dose in mid-afternoon. He said it never caused him to feel drowsy, but rather kept him awake at night if he had too much of it.[16] Dr Brooke, on his travels round Italy, met a Turkish doctor who had prescribed opium repeatedly for himself and for his patients. It was, he said, 'a gross mistake' to think that taking large quantities as a medicine would always develop into habit. He said much the same about the belief that addiction would shorten life. Opium was not only a sedative, it was 'a divine drug' that cured rheumatism, gout, coughing, headaches and numerous other disorders.[17]

This catalogue of medical applications was very much in line with therapeutic practice in Britain. Opium was already in general use by ordinary people as a remedy for pain, restlessness, coughing and spasms in a wide range of diseases.[18] During the eighteenth century, medical writers explored and debated the drug's value in specific complaints, continuing with Alston's experiments in the 1740s and culminating in the compendia of Crumpe and Leigh.[19] Some of the leading medical educators of the day approved the use of opium in particular conditions, and young Company doctors went out to India convinced that the drug was essential to their practice.[20]

The chief medical text for the Indian Ocean littoral had long been Bontius's *De Medicina Indorum*. First published in 1642 in Latin, the international language of the medical profession, it went into an English edition in 1769,[21] which suggests an expanding demand from a non-medical readership who needed to be able to treat themselves. Bontius was a champion of opium: 'without the assistance of opium and opiates, we never could, in these hot countries, cure either the dysentery, cholera, ardent fevers, or other bilious diseases'. He also prescribed those remedies for dropsy, hepatic flux, certain kinds of belly ache, herpes and spasms. An anonymous commentator who wrote footnotes for the English edition generally agreed and thought that even where Bontius's understanding of disease was faulty, 'in his ultimate recourse to opium, he proceeded upon the most rational and auspicious principle'.[22] By the end of the century, opium treatments were advocated by Company doctors in many situations. Francis Balfour, who practised for years in Calcutta and was an authority on fever, wrote that in serious cases 'there is no remedy so powerful and certain as opium'.[23] Balfour was also typical of his time in being alert to the value of native remedies. He popularised an ointment for ophthalmia, which consisted of butter, opium, alum and small portions of turmeric and neem leaves.[24] Within the European community in India, opium was self-administered for all sorts of complaints. The most celebrated example was Robert Clive who used the drug, sometimes in very large quantities, as a palliative for severe abdominal pain and gout.[25] Doctors in army barracks and public hospitals used opium to treat those twin scourges of life in India: malaria and dysentery.[26] The enormous doses which patients were able to tolerate showed that the drug's effects were not racially determined, but varied with the pathology of the individual.

The East India Company and opium control

How, then, did East India Company officials regard opium in the late eighteenth century, when they took control of the opium-producing areas of Bengal? They would have regarded it first and foremost as a medicine, a medicine that was absolutely vital to their survival in that unhealthy environment. They knew that Indians also used it as a medicine. Indeed, by relying on their servants for medical advice, as they often did, they reinforced their common interest in its curative effects. They recognised that Indians used it as a stimulant and a sedative and so did EIC officials; that it was regarded as an aphrodisiac, but they probably treated these claims with an appropriate degree of scepticism. They knew that it was eaten and drunk by Indians on ceremonial and recreational occasions and that it might be taken to excess and even to addiction. They saw nothing peculiarly 'oriental' in this; rather, as opium intoxication met with the same public disapproval as drunkenness in Britain, they made a distinction between harmless consumption and pernicious over-indulgence, as British society did with alcohol.

The first step in controlling the drug's use came from this equivalence between opium and alcohol. In the early 1790s, the Company's government in Bengal began to work out a settlement with the great landed magnates of the countryside, the zamindars. The key element in this arrangement was giving the zamindars the powers and responsibilities of landlords, thus identifying a relatively small group of men from whom the land revenue could be collected. In order to encourage the new landlord class to be progressive, the revenue demand was fixed in perpetuity, thus holding out to the zamindars the prospect of a rising return from agricultural development. In addition to rights over the land and its revenue, the zamindars had long control over village and urban markets, from which they extracted a motley assortment of rents, taxes, tolls and bribes. The Company saw the taxes and tolls as a restraint on trade, something that obviously could not be tolerated, and in a series of measures swept them away. In one or two cases, however, it transferred the right of taxation to itself. Chief among these was the tax on stills making spirituous liquors. Soon after the order for this was issued, the Collector of Bihar wrote to the Board of Revenue and pointed out that the immoderate use of toddy (fermented palm juice) was 'more destructive to the health and morals of the lower classes . . . than even Arrack' and the spirit of the order was applicable to this also.[27] The Board then modified its order so that it covered intoxicating liquors generally. This raised the issue of drugs in the minds of other Collectors. One suggested that bhang (cannabis leaves) also fell within the scope of the rules; another sought advice on the disposal of an existing zamindari tax on 'a certain preparation of Opium smoked by the natives under the name of Muddut'.[28] In this way the scope of the Government's orders gradually expanded until, by the time they were codified under Regulation XXXIV of 1793, bhang, ganja (cannabis flowers), charas (resin) and other intoxicating drugs were subject to an annual tax, fixed by the Collector of the district, and their sale and manufacture were forbidden without a licence.

Madat ('muddut') was the first opiate to come under regulation and it was the one that attracted the greatest condemnation throughout British rule in the subcontinent. It was prepared by chopping up betel leaves, heating the pieces on hot earthenware so that they curled into tiny balls, soaking the balls in a solution of opium and putting them out to dry. The balls were placed one at a time in the funnel of a hookah and heated with a lighted straw while the smoker inhaled. The procedure was very similar to the smoking of chandu (opium syrup) in China, though the technique does not seem to have been borrowed from there. Betel was not the only vehicle for the opium as an extraordinary variety of substances was used, even cow dung. Unsurprisingly, the health of madat smokers was always poor. The reports suggest that madat smoking was the recreation of a deviant minority, made all the more reprehensible by the violation of caste norms that was involved in mixing with strangers in the smoking dens and putting one's lips to a hookah that had been used by others.

This minority was certainly small. Some officials pointed out that madat smokers were so few in number that the tax on licenses would not be worth the trouble of collecting. The drugs of choice in Bengal and Bihar were tobacco and alcohol. Tobacco was smoked and chewed in enormous quantities, in some districts by virtually the whole of the adult population, both male and female. Spirits were made from rice or mahuya flowers in simple village stills which were impossible for the authorities to locate. Huge quantities were made in a day and then drunk very rapidly, before they went sour in the heat. Francis Buchanan, who made a tour of the districts along the Ganges in 1810–1814, said of Bhagalpur: 'In no country have I seen so many drunk people . . . men, who from their dress were far above the vulgar, lying on the road perfectly stupified with drink, and that in the midst of day, and in places far removed from the luxury and dissipation of towns.'[29] In a few districts where cannabis was grown, ganja and bhang were popular, and were considered healthier than any other intoxicant. Buchanan numbered the abusers of opium in tens and hundreds, while the drinkers of spirits were numbered in hundreds of thousands.[30]

The Regulation of 1793 soon began to seem inadequate. The tax on spirits brought in handsome returns, more than Rs. 200,000 in Behar district alone by 1798, and, with the land tax fixed in perpetuity, the excise on intoxicants offered good prospects for enhancing government revenue.[31] Officials in Calcutta wanted to set uniform tax rates for the whole province instead of allowing Collectors to set them district by district. The excise returns showed how little progress had been made in checking the problem of drunkenness, something which had been a government aim from the start. They also showed that, in some parts of the province, the use of drugs was greater than what had first been thought, owing to the unexpected popularity of cannabis in its several forms. When it was pointed out that raising the tax on drugs might put some shopkeepers out of business altogether, since they were almost as impoverished as their clients, the government did not flinch; if a high tax resulted in 'a total prohibition, the loss of a trifling Revenue will be esteemed by us as an object of little importance, considered with the general Interests of the Community'.[32] The Revenue Board was instructed to canvass the opinions of local officials and draft a new regulation specifically for drugs.

British officials were naturally inclined to condemn all forms of insobriety, and when asked to consider drugs apart from the general pattern of intoxication in the community, they saw these substances as an evil of especially menacing proportions. They denounced one drug after another as 'highly pernicious', 'extremely prejudicial' or 'violently intoxicating'.[33] Madat was universally condemned, usually for causing stupefaction and debility rather than for anything more socially disruptive. Ganja and charas were said to produce 'immediate intoxication', embolden criminals and lead to insanity, and even bhang, which had medical uses in most of the districts, did not escape criticism. One or two of the Collectors reviewed the evidence with a philosophical shrug and said there was no hope of effective prohibition; the rest called for heavy taxation or an outright ban. Opinions also differed in Calcutta, the Governor General's council ultimately taking a more severe position than the Board of Revenue. The new law (Regulation VI of 1800) required shopkeepers selling drugs to take out a licence, pay a monthly tax and keep order on their premises. Madat and charas, 'being of a most noxious quality, and... always highly prejudicial and dangerous to health', were banned entirely from sale or manufacture.[34]

Pure opium, as distinct from madat, was excluded from these developments on the orders of the Governor General. Two considerations lay behind this omission, the first being the medical impact of a rise in opium's retail price. The government had recently been advised by its Inspector of Drugs that 'Opium is a medicine of so much importance the diseases in which it is most frequently prescribed are so dangerous, and the quantity in which it is exhibited is so small' that consumers would seek out the best, whatever its cost.[35] The assumption was that a small proportion of the export opium from the Calcutta auctions would be sent inland and retailed to native druggists. However, the government's monopoly of opium production already acted to push up the auction price, and officials thought it would be wrong to add taxes to the retail sale. It did not matter if ganja and bhang were taxed out of existence, but opium was a different matter.

The medical importance of opium was confirmed by reports from the Bengal districts. During the consultation that preceded Regulation VI, most officials commented on opium in its pure form, even though they were not required to do so. Some of them simply said that its properties were well known, others said that it was widely and continually used as a medicine, and some went on to note its potential for addiction if taken to excess. Only in two of the districts, Mymensing and Sylhet, both of which bordered Assam, was opium consumption thought to constitute a significant social problem. Here the opium was impregnated into rags and traded under the name of koppah. In Sylhet many people carried the rags about and chewed them incessantly, preferring this form of intoxication to ganja or bhang. In Mymensing the Muslim population soaked the rags in water and drank the solution. The new Regulation bracketed koppah with madat and charas as a drug that deserved a total ban.

A second factor that contributed to the separation of opium from other intoxicating drugs was the reform of the system of opium production and manufacture that was under way in the late 1790s. The government's aim was to eradicate poppy cultivation in Bengal itself, notably in the districts of Rangpur and Bhagalpur, and concentrate production in Behar and Benares, where the opium was of better quality and the peasants could more easily be supervised. Under Regulation VI of 1799, the production and trafficking of non-Company opium became a criminal offence. This policy, which was aimed primarily at the protection of the export trade with China, had obvious implications for the supply to local consumers. As the policy began to bite, and as it was applied to new Company territories further up the Ganges valley, opium became much more difficult for Indian consumers to obtain.

During the next 16 years, as opium eaters struggled to get their supplies, officials argued with each other over where the proper balance lay between supplying opium to the sick and withholding it from the addict. In the 1790s the opium agents in Patna and Benares were given permission to send 'any reasonable quantity' to a few selected shops in areas where the lack of opium would have been 'a hardship'.[36] These shops were seen as a convenient way of disposing of confiscated or adulterated opium, and the poor quality of their stocks attracted few customers, which gave the impression that demand was small. However, John Fombelle, the Magistrate of Bhagalpur, reported a very different position in his district. He said that to those in the habit of using it, opium was 'as necessary as the air they breathe'.[37] They were finding it very difficult to get supplies because local druggists thought illicit opium from the villages would be indistinguishable from Company opium from Calcutta and the laws against the one would inevitably operate against the other. Fombelle's superiors were complacent about this state of affairs as 'it is by no means the wish of Government to encourage the consumption of opium'. They decided that 'should such an idea prevail [that] restrictions do exist [the government] does not think it at all desirable that any steps should be taken to remove it'.[38] Fombelle's riposte was to forward a pathetic letter from 'respectable' residents of Bhagalpur town. The signatories, confirmed opium-eaters and mostly Muslim, bewailed the disappearance of opium from the market for without it they could find 'neither a relish for life, nor the means of mitigating the pangs of death'.[39] Officials at the Board of Trade in Calcutta were moved very little by these entreaties and not at all by Fombelle's suggestion that opium should be retailed under the same restrictions as other drugs. In their view, opium was used so infrequently in Company territories that a shop here and there, supplied from Patna or Benares, was a perfectly adequate supply; anything more would encourage the population in this 'most fascinating' of luxuries, until it was as generally used as it was in China, with 'consequences fatal to the morals, and the health of the inhabitants'.[40] The Board of Trade argued that there was nothing to prevent buyers at the auctions from sending opium back inland if there was a genuine demand for it; however, in reality the auction price, the organisation of the Calcutta market and the threatening terms of the Regulation of 1799 all militated against this. In 1807, another survey of district opinion, provided this time by magistrates, led to a total of 10 shops being approved for the entire population of Bengal and the new Gangetic provinces: seven in the old opium-producing districts of north Bengal, and three in the district of Kanpur (Cawnpore).[41]

During the following years the upward pressure on supplies continued, with the Board of Trade gradually giving ground to critics in Calcutta and Collectors in the districts. More petitions were received and more shops opened on an ad hoc basis. In 1813 the government (by Regulation X of that year) at last brought opium into line with other intoxicating drugs. Collectors were to apply to the Board of Trade or the opium agents in Patna and Benares for enough opium to meet the medical needs of their districts, selling it through licensed drug dealers in the principal towns. Under the new Regulation, Collectors were required 'to discourage to the utmost extent of their means the sale and consumption of opium, except for medicinal purposes'.[42]

The more recognition the government gave to opium consumption, the greater that consumption was seen to be. Within a few months it was clear that neither the Board of Trade nor the opium agents could provide an adequate supply.[43] In one district, a collector even told his local drug dealers to buy illicit opium rather than allow supplies to dry up altogether.[44] Smuggling was widespread. Opium came from tiny poppy beds in countless villages, floated down rivers in boatmen's cargo, concealed in the saddlebags of bullocks on jungle trails, and carried by pilgrims into every temple complex. The Collector of Hijili suddenly discovered through a tip-off that he had 25 illicit dealers in his district town. A dawn raid, somewhat frustrated by the fact that his informer had changed sides and warned some of the dealers, yielded a significant haul. A day later a throng of 200 people gathered outside the Collector's office and entreated him to release the impounded opium on which, they said, their very lives depended. The Collector was so moved by this that he sold them the opium himself.[45] He also began an enquiry into the true levels of consumption in his district, which showed that about 25 per cent of the population could be classed as opium-eaters, consuming a total of 21,000 lbs of the drug annually.[46] Similar estimates from Rangpur, which had been discounted by Calcutta officials for years, now began to look as if they might be true.[47]

Officials were also beginning to take a broader view of the therapeutic importance of opium. The drug obviously had to be classified as a medicine when it was the recognised treatment for a particular disease. For example, in the marshy riverine tracts of central Bengal, people dosed themselves with opium to treat colds and rheumatism; elsewhere in Bengal, and universally in the upper provinces, opium was used to treat bowel complaints of every kind.[48] A classification was more problematic when opium was taken on a daily basis to soothe the aches and pains of old age. If regular users had their supply disrupted and began to feel unwell through the onset of withdrawal symptoms, the essential remedy for their 'illness' was another dose. In some of the Company's territories it was normal to give opium to children as a treatment and prophylactic for the diseases of infancy and as a general tonic. In one way or another 'there is scarcely a family within this City or at Muttra that do not use it', reported the Collector of Agra.[49] The government was driven to concede that not only was the drug 'essential to the health of many classes of people' but that 'the attempts hitherto made to restrain the consumption strictly to what is required for medical purposes, that is for the use of persons actually under the influence of disease, had only served to aggravate the problem'.[50]

A resolution of these difficulties came with Regulation XIII of 1816 which codified and revised all the laws relating to opium. On the supply side, the government accepted that it would have to make opium available from its existing agencies in Patna and Benares, as well as opening a new one in Rangpur to revive poppy cultivation in that district and distribute opium to the neighbouring parts of Bengal. To regulate consumption, the new law looked forward to a network of shops, set up by the collectors and run by nominees earning a monthly salary or a commission on sales. The Board of Revenue was to fix the retail price and the tax payable by the vendor, pitching these at a level which would not provoke smuggling. Limits were placed on an individual's possession of opium and unauthorised selling was to be an offence, except in cases where opium was being administered as a medicine. However, the Collectors were no longer required to discourage consumption, that was now a function of the retail price, or to confine shops to the larger towns. The Regulation's aim was less restrictive and more consistent with the actual patterns of consumption; to limit opium use 'as far as possible, to cases in which it may be necessary or salutary'. The prohibition of madat was dropped, probably because it was seen to be futile.[51]

Conclusion

A quarter-century of legislative experiment had therefore brought the Company's government back to the point from which it had started. The medical uses of opium, broadly defined to include both self-medication and professional treatment, were of over-riding importance. Good quality supplies of the drug had to be made widely available. The system established in 1816 allowed Indians the continued use of one of the most valuable drugs in the domestic medicine chest, a simple means of soothing body and mind, of controlling pain and alleviating symptoms. It replicated the central role of opium in British medicine in India during the nineteenth century.[52]

Why, then, did the Company's policy temporarily lose its way? Commercial and administrative concerns encouraged the Company to clear away the tangle of local taxes left over from zamindari control of the countryside. The hope of financial advantage prompted it to adopt the tax on spirits as one of its own revenue measures. This led inexorably into a consideration of the use of intoxicants generally. Opium was actually the least problematic of these, apart from tobacco. The most widely used, and probably the most socially disruptive, was alcohol; the most novel, and apparently the most dangerous to the health of the individual, was cannabis. The Company's new-found concern about drugs was driven by cannabis. Madat was too small a problem to cause much anxiety.

The Company's paternalistic concern about drugs was undoubtedly genuine. It was reiterated over and over again, up and down the hierarchy of government, in correspondence that had no value as publicity. It was an extension of the disapproval of drunkenness in British culture, and the use of the term 'intoxicating drugs' showed how British officials approached the subject in India. There is no indication that the Company promoted the use of drugs in India as a source of profit for itself. The fact that it ultimately resorted to a market mechanism to control them was due to the impracticality of any other form of interference in the daily life of the inhabitants. 'In no country it is probable would such an endeavour on the part of the Government to controul [sic] the habits of the people succeed', wrote one official, 'and it is needless to state how little suitable to the British Government of India would be the minute controul which would be requisite to give it here any prospect of success.'[53]

The main obstacle to a properly balanced policy on opium consumption was the export trade to China. Everything about it conspired against the Indian opium-eater as increasing quantities of the drug had to be supplied to the auctions in Calcutta, production had to be concentrated around Patna and Benares so that quality could be maintained and the manufacture and sale of opium elsewhere had to be restricted in case it led to competition in the export market. For many years the officials in the Board of Trade in Calcutta, who had responsibility for exports, held a veto over opium policy as a whole. Petitioning from the districts, humanitarian lobbying from the Board of Revenue and a growing appreciation of the drug's medical ramifications eventually forced the compromise that passed into law as Regulation XIII.

By 1816, opinions about the opium-eater had softened considerably from the censorious tone adopted towards all drug takers in the late 1790s. Those who slid from therapy into dependency were objects of pity. Though the proportion of opium-eaters in the population was probably quite high, there is little sense in the records of the time that this was damaging to society as a whole and constituted a problem that the government ought to attack. Somewhere between therapy and dependency there was a level at which opium was taken as a daily tonic, and officials could see little harm, and possibly much good, in allowing this to continue.

There was, then, a significant convergence of British and Indian views on the subject of opium use. There may also have been similar attitudes to its abuse. Madat smoking was so furtive and so comparatively rare that few, if any, of the British officials at the time would have had first-hand knowledge of it. Their information would have been based on the evidence and opinions of the 'respectable natives' that they normally consulted on such occasions. As Indian society regarded medicinal opium-taking by mouth as normal and valuable, but saw the recreational smoking of opiates as a perversion, it is very likely that British people in India made the same distinction. It is also possible that this enabled the British to live with the double standards arising from their opium trade with China. If the Chinese were apparently perverse enough to smoke the drug as a recreation instead of taking it by mouth as a medicine, the British would have seen no need to regret any addiction that followed. Thus an Englishman in Calcutta, who started the day with a pill of opium to settle his stomach, and ended it by drawing deeply on the tobacco in his hookah, could calmly distance himself, and the society around him from the peculiar oriental perversions that he saw still further to the East.

Notes

1. J. Grose, *A Voyage to the East Indies*, volume 1, second edition (London: Hooper and Morley 1766), p. 122.
2. T. Pires, *The Suma Oriental of Tome Pires. An Account of the East, from the Red Sea to Japan, Written in Malacca and India in 1512–1515* (London:

Hakluyt Society 1944), p. 513; G. Da Orta (translated by Sir C. Markham), *Colloquies on the Simples and Drugs of India* (London: H. Sotheran 1913), p. 330; D. Barbosa, *A Description of the Coasts of East Africa and Malabar in the beginning of the Sixteenth Century* (London: Hakluyt Society 1866), p. 57.
3. J.H. Linschoten, *John Huighen van Linschoten his discours of voyages into ye Easte and West Indies* (Amsterdam and Norwood, NJ: Facsimile edition of the English translation of 1598, 1974), p. 124.
4. W. Foster, *Early Travels in India 1583–1619* (Delhi: S. Chand reprinted 1968), p. 71.
5. Ibid., p. 116.
6. F. Bernier, *Travels in the Mogul Empire A.D. 1656–1668* (Delhi: S. Chand reprinted 1968), pp. 39–40; A.P. Hove, *Tours for Scientific and Economical Research made in Guzerat, Kattiawar and the Conkuns in 1787–88* (Bombay: Gibson 1855), p. 34.
7. Grose, *A Voyage to the East Indies*, pp. 119–20. Similar reports come from Persia and Turkey.
8. J. Kerr, 'The Culture of the White Poppy and the Preparation of Opium in the Province of Bihar', OIOC, Mss.Eur.K. 45.
9. J.Z. Bowers and R.W. Carrubba, 'Drug Abuse and Sexual Binding Spells in Seventeenth Century Asia: Essays from the Amoenitatum Exoticarum of Englebert Kaempfer' in *Journal of the History of Medicine*, 33, 1978, pp. 329–36.
10. Sir T. Herbert, *Some Years Travels into Divers Parts of Africa, and Asia the Great. Describing More particularly the Empires of Persia and Industan* (London: Everingham *et al.* 1677), p. 311.
11. R. Mead, *A Mechanical Account of Poisons; in Several Essays* (London: Ralph South 1702), p. 143.
12. Ibid., p. 120.
13. D. Campbell, *A Narrative of the Extraordinary Adventures... of Donald Campbell, Esq. Of Barbreck* (London: Tiebout 1798), pp. 150–1.
14. A. Russell, *The Natural History of Aleppo and Parts Adjacent... Together with an Account of the Climate, Inhabitants and Diseases* (London: Miller 1756), pp. 83–4; F. Hasselquist, *Voyages and Travels in the Levant, in the Years 1749–52, Containing Observations in Natural History, Physick, Agriculture and Commerce* (London: Davies and Reymers 1766), p. 299.
15. J. Dallaway, *Constantinople Ancient and Modern* (London: T. Bensley 1797), p. 78.
16. E. Smyth, 'Of the Use of Opium Among the Turks' in *Philosophical Transactions*, XIX, 1798, pp. 288–90.
17. N. Brooke, *Observations on the Manners and Customs of Italy... Likewise an Account of Cures Produced by a Preparation of Opium, in a Variety of Obstinate Cases, According to the Practice in Asia* (Bath: Cruttwell 1798), pp. 36–40.
18. W. Buchan, *Domestic Medicine; or, the Family Physician* (London: Strahan 1779), passim.
19. C. Alston, 'A Dissertation on Opium' in *Medical Essays and Observations*, volume V, pt.1, 1742, pp. 110–76; S. Crumpe, *An Enquiry into the Nature and Properties of Opium; Wherein Its Component Principles, Mode of Operation, and Use or Abuse in Particular Diseases, are Experimentally Investigated* (London:

Robinson 1793); J. Leigh, *An Experimental Inquiry into the Properties of Opium* (Edinburgh: Eliot and Robinson 1786).
20. For a discussion of this, see my 'Opium as a Medicine in Nineteenth-Century India' in S. Bhattacharya (ed.), *Imperialism, Medicine and South Asia (1800–1950)* (forthcoming).
21. J. Bontius, *An Account of the Diseases, Natural History & Medicines of the East Indies*, translated from the Latin with annotations by a physician (London: Noteman 1769).
22. Ibid., p. 173.
23. F. Balfour, *A treatise on putrid intestinal remitting fevers* (Edinburgh: William Smellie 1790), p. 126.
24. F. Balfour, *A treatise on the Action of Sol-Lunar Influence* (Edinburgh: William Smellie 1791), pp. 84–5.
25. M.B. Jones, *Clive of India* (London: Constable 1974), pp. 71, 191, 241 and 299. For a more mundane example, see J. Gilchrist, *A Collection of Dialogues... on the Most Familiar and Useful Subjects* (Calcutta: 1804), pp. 204–5.
26. J. McGregor, 'A Memoir on the State of Health of the 88th Regiment... as presented to the Medical Board, Bombay' in *Edinburgh Medical and Surgical Journal*, I, 1805, pp. 266–89; N. Chevers, *A Commentary on the Diseases of India* (London: Churchill 1886), p. 158.
27. A. Seton, Behar, two letters, n.d., Bengal Board of Revenue Consultations (hereafter BBRC), 13 and 29 October 1790. The Collector was the senior British revenue officer in each district.
28. H. Colebrooke, Purnea, 11 May 1792, BBRC, 18 July 1792; Collector, Rangpur, 26 May 1791, BBRC, 14 October 1791.
29. 'An account of the district of Bhagalpur' [1811], p. 44, OIOC, Mss.Eur.D81.
30. Ibid.
31. Summary table, dated 20 September 1799, in BBRC, 18 October 1799.
32. Governor General in Council to Board of Revenue, 2 March 1798, Bengal Revenue Consultations (hereafter Rev.Cons.), 2 March 1798.
33. Reports from collectors, BBRC, May 1798.
34. *Regulations* (n.p., n.d.), OIOC, V/27/120/57A.
35. J. Fleming, Inspector of Drugs, to Board of Trade, 12 January 1794, Rev.Cons., 1 June 1794, p. 3.
36. Board of Trade to Governor General in Council, 13 April 1798, Rev.Cons., 27 April 1798, p. 2.
37. Fombelle to Secretary, Revenue and Judicial Dept., 7 October 1800, Bengal Judicial Consultations, 16 October 1800, p. 3.
38. Ibid., p. 4.
39. Copy with letter from J. Fombelle, magistrate, Bhagalpur to Bengal Revenue Dept., 18 February 1801, Rev.Cons., 5 March 1801. The situation had worsened after the eradication of poppy cultivation in Bhagalpur, because local opium-eaters could no longer salvage opium from the rags and pots discarded by the cultivators.
40. Board of Trade to Vice-President in Council, 27 November 1801, Rev.Cons. 14 January 1802, p. 3.
41. Rev.Cons., 21 August 1807, pp. 1–30.
42. 'Brief Historical Sketch of the Law and Rules regarding the Supply and Sale of Excise Opium in Lower Bengal' by K.G. Gupta, 'Royal Commission on

Opium', *Minutes of Evidence taken before the Royal Commission on Opium between 18th November and 29th December 1893; with Appendices*, volume II (London: 1894), Appendix XXII, p. 426.

43. D. Campbell, Opium Agent, Patna, to Board of Trade, 18 December 1813, Rev.Cons., 29 April 1814, p. 45.
44. R. Vansittart, Midnapore, to Bengal Board of Revenue, 3 October 1814, BBRC, 15 November 1814, p. 25. One of the dealers was arrested by the police and charged by the district magistrate, leading to an embarrassing clash between the authorities. The Board of Revenue said it could not sanction the sale of contraband opium, but neither did it reprimand the collector.
45. C.R. Crommelin, Hijili, to Board of Revenue, 26 July 1814, Rev.Cons., 17 September 1814, p. 20. This time the Board endorsed the breach of the regulations. Hijili was the name then given to the coastal areas of Midnapore.
46. Ibid.
47. Board of Trade to Gov.Gen. in Council, 23 July 1807, Rev.Cons., 21 August 1807, p. 1.
48. M. Rennell, collector, Dacca to Board of Revenue, 14 June 1814, BBRC, 8 July 1814, p. 43; W.L. MacGregor, *Practical Observations on the Principal Diseases . . . in the North-Western Provinces of India* (Calcutta: Thacker 1843), p. 231.
49. A. Wright, Agra, to Board of Commissioners, Farrukhabad, 14 May 1815, Rev.Cons., 10 May 1816, p. 5.
50. H. Mackenzie, secretary to Revenue Department, to Board of Revenue, 9 August 1816, Rev.Cons., 9 August 1816, p. 16.
51. *An Abstract of the Regulations of Government . . .*, volume IV (Calcutta: Thacker 1828), pp. 134–58; see also drafts in Rev.Cons, 15 March 1816, 7 and 10 May 1815, p. 7.
52. See my 'Opium as a Medicine'.
53. H. Mackenzie, Revenue Department, to Board of Revenue, 9 August 1816, Rev.Cons., 9 August 1816, p. 16.

5

'Cannot We Induce the People of England to Eat Opium?' The Moral Economy of Opium in Colonial India

John F. Richards

Introduction

Most historians who have written about opium in the nineteenth century have deplored in no uncertain terms India's role in the world opium trade. They have been dismayed that from the late eighteenth to the early twentieth century the Indian subcontinent was the primary source of raw opium produced for world market demand. They have seen Indian profits from opium as one of the primary reasons for Britain's aggression against China in two wars that prevented the Qing dynasty from successfully banning imports of Indian opium. In my view, this condemnation, heavily freighted with moral judgments, derives from the demonization of opiates by the rhetorical excesses of the world drug wars of today. Because of these presentist blinders, historians have too willingly accepted the arguments of generations of British and American opium reformers who attacked the Indian Government for its policies.

Generally historians have not confronted the cultural imperialism of those Western opium reformers who ignored the innocuous role opium played in Indian society itself. Because of their anti-drug bias, historians have ignored the medical benefits of opium use. They have overlooked the benign social role that circumscribed opium use played in some regions and within some groups in Indian society. Deeply influenced by current drug war rhetoric, they have branded as evil and criminal the economic benefits the opium trade conferred upon the colonial economy and Indian society. In fact, as a cash crop and a valuable export, opium benefited India. The economy of opium was moral, not immoral.

This presentist anti-drug attitude permeates M. Emdad-ul Haq's *Drugs in South Asia: From the Opium Trade to the Present Day*, a recent comprehensive study of opium and drugs in South Asia. In its opening pages, Professor Haq makes the extraordinary claim 'that the British policy for the maximization of revenue from drugs largely helped in the promotion of drug addiction and drug trade in South Asia'.[1] Similar views color the interpretation of Carl Trocki in his stimulating *Opium, Empire and the Global Political Economy*. Trocki deplores the 'drug plague' foisted on China. He revives an earlier phrase by John Fairbank to characterize the nineteenth-century opium trade as 'the most long-continued and systematic international crime'.[2] These attitudes still distort the historical interpretation of the role played by opium in Asia during the nineteenth century. It is past time for a more balanced, revisionist approach that does not apply the automatic reflexive stereotypes of the twenty-first century drug war to the study of opium.

Assessing opium production in India

In the topsy-turvy world of drug history, what would ordinarily be viewed as commendable achievements become immoral activities. Historians condemn the Indian government for its success in developing opium as one of India's most valuable exports and for managing to tap those exports to generate substantial revenues for the state. The colonial state collected opium revenues in three ways: profits from the state-controlled monopoly of Bengal customs duties (pass fees) levied on chests of Malwa opium produced within the princely states of western India and exported from Bombay; and finally excise returns from domestic sales of opium via a network of licensed retail shops. Throughout the nineteenth century, opium revenues were one of the most significant sources of funds available to the British colonial regime in India. Over the half-century between 1840 and 1889, total opium revenues averaged 6.6 million rupees per year or an average of 15 per cent of total Indian revenues.[3] During the same period, opium exports from Calcutta and Bombay averaged well over one-fifth in value of all products shipped from British Indian ports.[4]

Moreover the monies collected by the state from the Bengal monopoly, the Malwa pass fees and the sale of opium to domestic excise shops had little effect upon the Indian subjects of British rule. Unlike the salt tax, opium was a benign source of revenue that hardly affected the average Indian subject of the British Raj. The official monopoly over the production and sale of salt with its high monopoly prices was a cruelly regressive tax. Despite its condemnation by successive generations of reformers and officials, high-priced salt continued to demand a disproportionate amount of ordinary Indian household budgets. Unlike opium, salt was an essential item in the Indian diet. Unlike opium, salt was not an export. Instead the Indian government even imported expensive British salt to compensate for shortfalls in Indian production.

High demand for opium made poppy cultivation a profitable activity for peasant farmers in the eastern Gangetic plain and in western India. Under the Bengal monopoly the British state in India created and fine-tuned an official monopoly over poppy cultivation and opium processing during the first 40 years after the conquest of Bengal. By the mid- to late nineteenth century, the colonial state had marked out a well-defined region in which well over a million peasant cultivators granted yearly licenses could grow poppy on specified plots of land. The Indian Opium Department offered interest-free advances to these cultivators as well as a fixed price for raw opium at harvest. Under law opium cultivators were free to accept or to reject opium advances. Any sale or disposal of opium to private parties was illegal. The Opium Department processed the raw opium at two industrial establishments at Ghazipur and Banaras and sent it in kilogram-sized balls, 40 in a wooden chest, to Calcutta. There at monthly auctions, exporting firms and speculators bid on lots of chests to be exported either to the China market or to Southeast Asia.

Throughout the nineteenth century, Indian and British anti-opium reformers argued the Bengal system exploited peasant growers by forcing them to grow the crop despite inadequate prices paid by the state. To what extent, if any, coercive measures were used is difficult to determine, but in my view a systematic state policy that permitted coercion did not exist. Certainly the legal issue was clear: poppy growers had to fulfill the terms of their annual license, but they could elect not to accept advances and a license at any time. It is doubtful that the Opium Sub-Agents exerted overt pressure on individual peasant growers. It is possible that village officers or landlords demanded that those of their tenants, who were capable of doing so, grow poppy in order to obtain the highest possible rents. That is a question that still bears looking into by detailed empirical research.

What is certain is that in the latter decades of the nineteenth century, some 1.3 million peasant growers signed up to grow poppy on half a million acres of land. Year after year, the most proficient market gardeners in the eastern Gangetic valley, largely members of the Kachhi and Koiri subcastes, marked out small plots (less than one-half acre) of the best loamy village lands watered by wells to grow poppy.[5] Never do we see any of the sorts of resistance that marked the revolt of tenant indigo growers cultivating for British and Indian planters. Every year the Opium Department advanced 5–6 million rupees to growers. Every year the Department paid out around 18–19 million rupees for the final product. The price set by the state seems to have been sufficiently high to permit these growers who primarily used their own and family labor to obtain reasonable returns. Undoubtedly, interest-free advances combined with a guaranteed price made opium an unusually attractive crop, even though the Opium Department set the price at a finely calculated level aimed at reserving the greatest profits to itself. The Department also gave out interest-free advances so that growers could dig or improve wells. Each grower counted on additional returns from the drained poppy juice and steamed poppy petals sold to the Department; from seed beyond next year's needs which could be crushed to generate a light cooking oil and cakes of crushed seeds to feed to cattle; and not least, from small amounts of opium retained for home use or sold illegally (excise sales of opium in the opium-growing districts were far less than those outside this region). The entire package of returns for poppy growers was such that the most skilled cultivators engaged willingly. This is in sharp contrast to the combination of force and indebtedness that indigo planters brought to bear on their tenants to pressure them to grow indigo.

There have also been recurring suggestions that poppy growing contributed directly to food scarcities and famine in those years when the monsoon rains failed. Since the total area under poppy was less than 2 per cent of the entire area cultivated within the eastern Gangetic valley, it is hard to see how opium could have substantially contributed to food scarcities that stretched across vast regions of the subcontinent.[6] Beyond the eastern Gangetic plain the state imposed an ever-widening ban on poppy cultivation.[7] Bengal opium from the eastern Gangetic plain was only a part of the story, however. Western Indian or Malwa opium contributed nearly as much to the overall economy of the Indian Empire. Prohibited in British Indian districts, poppy growing flourished in some 90 Indian princely states in western India, notably Mewar, Bhopal, Jaipur, Marwar and Gwalior. Each of these dependent regimes obtained higher revenue yields from opium lands. Unlike Bengal, this was private enterprise in which Indian traders and merchants advanced funds to village moneylenders and headmen who in turn financed several hundred thousand cultivators each year to grow poppy. Traders processed and packed balls of Malwa opium and brought them to one of the ten British-run stations where they paid the set Bombay pass fee, obtained documents and sent the chests on to official warehouses at the port.

Unlike the Bengal monopoly, under the Malwa system the gross revenues from opium sales at Bombay remained in Indian hands (with the exception of the pass fees collected by the British). Total returns from Bombay sales, after deducting the pass fees, averaged between 31 and 32 million rupees per year between 1865 and 1885. In the next decade, average returns dropped to 19 million rupees and fell even more precipitously at the turn of the century.[8] This was a substantial infusion of funds, largely silver bullion, into the economy of western India. Cultivators, village moneylenders, haulers, opium brokers and sales agents all profited from this export industry. Moreover, many Indian trading families, including Parsis and Marwaris, took further profits by shipping Malwa opium to China and southeast Asia.

Curiously, however, the most recent study of Malwa opium by Amar Farooqui concludes that poppy cultivators in western India did not really share in the massive opium profits. Farooqui argues 'as in the Ganga [Bengal] region so also in Malwa, production of opium could hardly be regarded as a very profitable enterprise for the cultivator'.[9] He points out that the early British administrator John Malcolm calculated that in an average year the peasant's profit on five bighas of poppy would only be 1 rupee and 10 annas, and therefore 'the labour put into opium farming [was] so pathetic'. Instead, various intermediaries ranging from village moneylenders to landlords really soaked up any profits that returned to the countryside. Apparently then, both under the Bengal state monopoly and under the Malwa private enterprise system, poppy cultivators were both economically disadvantaged and oppressed. If that were true, one is forced to ask whether peasants growing any cash crop in either British or princely India gained any advantage from that activity. In other words, the perceived oppression of the poppy cultivator had little to do with whether the crop was opium, sugar, wheat, rice or any other crop grown for market. This conclusion in turn leads us to the convoluted, much-debated controversies surrounding such issues as population growth, extended cultivation, land scarcity and rising prices and land revenue trends.

Assessing opium consumption in India

Throughout both British India and princely India, Indian consumers could freely buy and consume opium in its various forms. In British

India, the state licensed retail shops to sell Bengal and, sometimes, Malwa opium through excise shops at prices determined by the state. Revenues from licenses and the wholesale sales of opium formed a growing part of the figures reported for opium revenues each year. Always conscious of possible criticism from British reformers, the colonial administration tried to calibrate the number and distribution of retail shops, the amounts of opium made available and the prices charged so as to maximize revenue and limit consumption. Those officials making these calculations were always mindful of limits imposed by existing black-markets. If shops and supplies were too few and official prices too high, the incentives for illegal sales and smuggling of opium grew accordingly. In 1893 the excise shops selling opium totaled only 10,118 for all of British India for a ratio of one shop to every 21,000 Indians. The state also set limits to a single sale at retail (fairly generous at between 300 and 900 grains of crude opium) and possession of larger amounts was grounds for arrest.

If selling opium to the Chinese was evil, then unquestionably any sales of opium by the Indian government to its own people was also immoral, a view echoed by later historians. British opium reformers agitated for nothing less than the complete prohibition of opium production and consumption in India. In the early 1890s, they placed their hopes on the Royal Commission on Opium. To their dismay, the Commission resisted the anti-opium rhetoric and stated instead that opium use in India was at worst, benign and at best, positively beneficial.[10] 'The gloomy descriptions presented to British audiences of extensive moral and physical degradation by opium, have not been accepted by the witnesses representing the people of India.'[11] According to the Commission, the people of India who consumed opium used it moderately and judiciously. Much consumption was for valid medical reasons in which Indian physicians or Indians themselves took opium to relieve pain, ease intestinal problems caused by dysentery and other similar diseases and to control severe, recurring coughs. They gave small amounts to infants to ease discomfort and permit sleep. They took opium as a work drug to ease the pangs of hunger, cold, discomfort, arduous physical labor. They took opium as a relaxant and mood-altering drug after work or on festival occasions.

In fact, for most Indians the attacks by British reformers on Asian use of opium was ironic. The opium reformers offered culturally biased judgments as to which mood-altering drugs were desirable. Most Indians regarded the pervasive use of alcohol in British society as far more repellent and harmful than the use of opium or cannabis throughout the subcontinent. They regarded the spread of beer or spirits drinking as a colonial habit that was being forced upon India by the addictions of their colonial rulers. They worried that if opium or cannabis were restricted further, then this would lead to greater use of alcohol. Witness after witness to the Indian Opium Commission made this comparison. Among the most cutting were the comments of T.N. Mukherji, a Government of Indian official, who had actually visited Britain to help organize the Colonial and Indian Exhibition and the Glasgow Exhibition. Expressing his cynicism about the motives of anti-opium reformers, Mukherji asserted that many Indians thought that the current campaign was simply a ploy to sell more British beer and spirits in India. He quoted from an essay he had published in 1890:

> Cannot we induce the people of England to eat opium instead of annually spending more than two hundred crores of rupees in the consumption of alcoholic liquors? Opium is amazingly cheap, duty included; it prolongs life after a certain age, and it can be asserted with all the force of truth and seriousness that its substitution in place of alcohol... will bring back happiness to thousands of families in Great Britain and Ireland where there is no happiness now.[12]

Conclusion

Mukherji was absolutely right. The anti-opium campaign in Britain and in the United States emerged from a profound cultural bias that judged opium evil primarily because it was Asiatic. Opium ingestion in India and even opium smoking in China if examined dispassionately was no more harmful to individuals or society at large than alcohol in western societies. If one drug was evil, the other must also be evil. Unfortunately, the cultural imperialism of the opium reformers had remarkable staying power. By the twentieth century, this judgment even pervaded the views of Asian nationalists, who by the end of the century saw opium use as a sign of inferiority. Generation after generation of historians who have written about opium and its derivatives have accepted the fixed ideas of what has become a world drug war with its own ideology, icons and victims. This chapter has attempted to summarize some of the historical evidence that undermines this set of discourses; opium production and marketing brought benefits to Indians involved at all points in the trade and to the South Asian economy as a whole, and its consumption in India in the nineteenth century was carefully controlled both by state and by social laws that ensured that a useful medicine and enjoyable intoxicant was available to those who chose to use it.

Notes

1. M. Emdad-ul Haq, *Drugs in South Asia: From the Opium Trade to the Present Day* (New York: Palgrave 2000), p. 1.
2. Carl A. Trocki, *Opium and Empire: Chinese Society in Colonial Singapore, 1800–1910*, Asia, east by south (Ithaca: Cornell University Press 1990). This is the title of Chapter 8.
3. J.F. Richards, 'The Indian Empire and Peasant Production of Opium in the Nineteenth Century' in *Modern Asian Studies*, 15, 1981, p. 161.
4. Ibid. Calculated from decennial averages in Table 2, pp. 166–67.
5. Ibid., pp. 59–82.
6. Richards, 'The Indian Empire and Peasant Production of Opium', 1981.
7. John F. Richards, 'The Royal Commission on Opium of 1895' in *Modern Asian Studies*, 36, 2002.
8. John F. Richards, 'The Opium Industry in British India' in *Indian Economic and Social History Review*, 2–3, 2002, pp. 176–7.
9. A. Farooqui, *Smuggling as Subversion: Colonialism, Indian Merchants and the Politics of Opium* (Delhi: New Age 1998), p. 66.
10. Richards, 'Opium Commission', p. 378.
11. Ibid.
12. Ibid., p. 420.

Part II
Control

6
Opium and the Trading World of Western India in the Early Nineteenth Century

Amar Farooqui

Introduction

For nearly half a century from about 1790 to the First Opium War, Malwa opium exports to Southeast Asia increased steadily; however, they provided little direct benefit for the East India Company throughout the period. Richards estimated that between 1821 and 1823, the average annual profit realised by the company from the sale of Malwa opium was only Rs. 7 lakhs. By 1832/1833, the company was earning about Rs. 20 lakhs from passes for 11,000 chests of Malwa opium, a mere 12 per cent of the total opium revenues of the Government of India.[1] As such the bulk of the surpluses which accrued from Malwa opium production were appropriated by non-company participants in the enterprise and Richards has observed that:

> Generating and collecting funds for opium advances to cultivators generated income for an array of village moneylenders, village headmen, and an ascending sequence of Indian bankers. Gathering raw opium, assembling, curing, packing and transporting it, employed thousands of labourers, guards, carters and other workers. As opium passed through a chain of transactions on its way to Bombay it offered reliable profits to numerous brokers, traders, and commission agents.[2]

This chapter will identify in greater detail the nature of the Indian groups that had stakes in, and profited from, the Malwa opium trade. This will act as a prelude to an attempt to quantify the impact of Malwa opium on India's colonial economy. It will posit that in Malwa, in contrast to the situation in Bengal, British attempts to dominate the opium trade failed, leaving it in the hands of local merchants. This involved the establishment of tortuous overland routes and complex financial transactions to keep the opium trade out of British hands. The strength of local control over the trade would ultimately have implications for local princely sovereignty, and the colonial politics with the British annexation of Sind and the First Opium War rooted in the frustration of the British officials at their lack of monetary reward from what should have been a lucrative trade.

Sindia as opium state

Of the numerous Indian states in Malwa at the beginning of the nineteenth century, all of which were subject to British control after 1818, Sindia was territorially the largest and politically the most dynamic. This was also the major opium-producing state in the region, even at a conservative estimate it was producing close to half the entire supply of Malwa opium at the time of the third Anglo-Maratha War.[3] This discussion will therefore focus on Sindia-ruled areas of Malwa.

It would be an oversimplification to regard the state as a puppet regime of the British or to portray it as being devoid of any independent authority after 1818. There were in fact three phases in the political history of the Sindia state between 1818 and the early 1840s. In the first phase the reigning king, Daulat Rao (1797–1827), attempted to preserve the structure of the state and much of its prestige from the settlement with the East India Company of 1818. It should be stressed that there was no military engagement between Sindia and the British during the third Anglo-Maratha War, and Daulat Rao had instead simply negotiated an arrangement with the British whereby he managed to retain a great deal of autonomy even though some of his authority was curtailed and Sindia territories were confined to Malwa. This phase lasted till Daulat Rao's death in 1827 following which one of his wives, Baiza Bai, became the regent of the state. Her regency (1827–1833) marks the second phase and it was in this period that the hitherto subdued anti-British sentiment at the darbar became increasingly conspicuous.[4] Baiza Bai having been deposed in a coup that was tacitly supported by the British resident in 1833, the third phase extends to 1843. The following decade was an unsettled one as conflicts between different factions of the Sindia army culminated in the ascendancy of a faction that favoured violent opposition to British interference. This outcome forced the East India Company to launch a full-scale military operation to subdue its opponents.

There can be no doubt that opium was one of the pillars that supported the Sindia economy during the 50 turbulent years after 1790. British supremacy after 1818 did not immediately deprive the Sindias of this support. Rather, smuggling networks developed to carry on a clandestine trade during the first two phases of the post-1818 era by taking advantage of the fact that the Company had virtually no say in day-to-day administrative matters. This was because in this period the power of the British Resident, usually the main instrument for colonial meddling in the internal affairs of Indian states that had accepted British paramountcy,[5] was so circumscribed in the case of the Sindia state that he was little more than a glorified newswriter until 1833. A number of factors explain this. In the first place, the resident was stationed at the Sindia capital in Gwalior which lay outside Malwa itself. As such he did not have much direct access to information regarding the region.

In addition, this interaction of successive Gwalior residents with the Sindia state did not go beyond formal exchanges at the level of the darbar. Here the resident was officially designated as no more than a *vakil* which in terms of court protocol placed him only on the same footing as the *vakils* or representatives of other indigenously ruled states in attendance (*mulazmat*) on the darbar. The routine appointment of vakils to various courts by rulers, chiefs and big zamindars was a common practice.[6] Under Baiza Bai even this interaction was reduced to a minimum. The resident was not permitted to have any dealings with the common people and had very little normal access to the darbar thus prompting one resident to remark that 'a more suspicious and distant durbar does not exist', while earlier, another resident, Stewart, had frankly admitted that 'we are very much in the dark as to what really takes place in the interior of the Palace'[7] Even a well-informed official like John Malcolm, who had far-reaching political and military clout and resided in the heart of Malwa, had only a vague notion of the Sindia state. In his famous 'Memoir of Central India', written shortly after the third Anglo-Maratha War, Malcolm dismissed the history of the Sindia state in just 25 pages while more than 180 pages was devoted to the Holkars of Indore and 65 pages to the minor political figure of Amir Khan.[8]

Negotiations for imposing on the Sindia state a treaty that would give to the East India Company exclusive rights to the opium produce of the state stalled because of this relative lack of British control over the Sindia elites allowing the state to avoid any firm commitment on the issue, especially during Baiza Bai's regency.[9] Opium was not just a question of revenues. Conceding a monopoly to the company was a serious infringement of the internal autonomy of the Sindia state. In a state that was so sensitive with regard to its formal authority and for which outward symbols associated with that authority were a matter of great concern, a monopoly was a privilege it could have reserved for itself alone.[10] In the case of opium it exercised no such monopoly. Nevertheless in a situation where there could be prolonged disputes over the smallest of ceremonial procedures, conceding a monopoly could scarcely have been looked upon as a minor affair.

The symbolic meaning of not granting to the company exclusive rights over opium should not be over emphasised. More important were the material interests that were involved. The Sindia ruling elite looked upon opium as a critical financial resource: opium helped the state to offset the losses that it suffered in the second and third Anglo-Maratha Wars. It is difficult otherwise to account for the relative stability of the state in the first half of the nineteenth century or its capacity to maintain a large army till 1844, estimated at 40,000 strong in Gwalior City alone in 1843–1844: moreover, one which could no longer be sustained through wars of conquest or raiding expeditions after 1818.[11] By the beginning of the nineteenth century the Sindia state was earning a large revenue from opium through customs duties, sales tax and the state's share of produce on land. It is not possible to quantify these items due to the paucity of source material.[12] However, it seems certain that there was a marked increase in land under poppy during the 1820s and 1830s.[13] This is reflected in the fivefold increase in shipments of Malwa opium to China between the 977 chests sent in 1818/1819 and the 5535 chests shipped in 1823/1824. This was followed by a 12-fold increase of 12,856 chests by 1830/1831. While the figure for 1818/1819 seems very low, because Dangerfield in 1820 had estimated that approximately 5000 chests of opium were being produced in Malwa at the time of the third Anglo-Maratha War, there is no denying the fact of a 100% increase, at the least, in opium production by the beginning of the 1830s. In 1832 the company issued passes for 11,000 chests, while 3007 chests were exported clandestinely from Daman in the same season. By the end of the 1830s, nearly 22,000 chests of Malwa opium were being shipped to China.[14] For estimations of potential revenue, nor should the quantity of opium that was consumed internally be ignored. This, according to Dangerfield, amounted to 2000 'Bengal maunds' of a total of 8692 *man* produced in Malwa.[15] Rajasthan seems to have been the main domestic market for Malwa opium. It seems clear that most of the increase took place as a result of extension of cultivation as there is no evidence for significant improvements in extraction of juice from poppy.[16] Poppy cultivation was part of the mixed cropping pattern of Malwa cultivators. In the early nineteenth century, dominant landed groups in the region coerced the peasants to set aside the bulk of the irrigated and high-fertility fields for opium cultivation, often at the expense of vegetables. The company's opium agent remarked in 1826 that 'the high price opium rose to, had induced the Ryots, or rather led the proprietors or renters of villages to compel the Ryots no matter how, whether by advances made through the sowcars, or assured advantages to sow every spot of ground they could with poppy to the exclusion of all the different kinds of vegetables, usually produced round villages'.[17]

The Sindia state had already put in place an elaborate infrastructure to realise duties on the transit and sale of opium in its territories. The collection of these duties was particularly systematic in an area like Mandsaur that was otherwise relatively backward as compared to, say, Ujjain. For example, in the second decade of the nineteenth century, there were 46 collection points for opium transit duties in the Mandsaur pargana. Of these, 24 were *nakas* or customs posts, while 22 were halting places or *khunts*, literally, places where bullocks could be unyoked and tethered.[18] One anna per *man* of opium was payable at each *naka*, while two annas per *man* had to be paid at each *khunt*.[19] Apart from the revenue which the drug yielded to the state and various intermediaries, there were the returns which investments in the opium enterprise brought to bankers, the landed aristocracy and members of the royal family. Some of these investments were of the nature of speculative advances to the state by *ijaradars* (revenue farmers) who then maximised their income by promoting opium cultivation. Leading *ijaradars* like Appa Gangadhar, who held the contract for opium-rich Mandsaur, earned profits by directly arranging for the supply of opium to wholesale markets.[20] Then there were the big banking firms that stood surety for the *ijaradars* or extended loans to the state against future revenues. The prospect of massive earnings from opium during the 1820s, when smuggling was at its peak, naturally increased the values of the revenue farms. The *ijaras* and personal landed estates should be viewed as captive sources of opium supply. This in turn provided to the ruling elite with an opportunity to amass huge private fortunes.

At the same time, sufficient cash reserves were available to the state: the official (royal) treasury of the Sindia state, called the *Ganga Jali*, reportedly had a reserve of 30 million rupees at the time of Daulat Rao's death. He is supposed to have 'prided himself upon never taking from, and always adding to, it'. Baiza Bai added another 10 million rupees to the treasury so that in 1833 Sindia reserves stood at 40 million.[21] A lengthy report on the administration of the Sindia state written in 1856 by S.C. Macpherson, the political agent at Gwalior, stated that up to 1833, 'Revenues were collected by a small number of Farmers' General, interested in the prosperity of their districts through holding permanent contracts; and by the three chief military commandants, to whom the more warlike tracts were entrusted. Those collector prefects exercised every function of civil administration, while they were nearly all, excepting the commandants, Mahratta Pundits, and, employed exclusively the agency of that clerkly and sacerdotal class, which... has ever dominated as a bureaucracy in each state.'[22] Such fiscal strength reinforced the stability of the ruling elite. Since many of the big *sahukars* were simultaneously contractors for recruitment of troops, there was a close link between opium profits and military mobilisation.[23]

It is quite likely that both Daulat Rao and Baiza Bai had invested part of their private wealth in opium. There is no direct evidence for this in contemporary colonial records but then the company's officials could not have known much about financial dealings at this level. Daulat Rao and Baiza Bai made their investments through the bankers at the darbar. Gokul Parakh was the leading Sindia banker during the 1820s and there are suggestions that Daulat Rao was probably a partner in Parakh's banking concern.[24] Parakh was also Sindia's Chief Minister and 'at the head of the mercantile body in Malwa', while the principal bankers of the Sindia state were his business associates.[25] Parakh dealt extensively in opium and at one stage was suspected by the company's officials to be instrumental in manipulating the prices at which peasants sold their produce for the Bombay auctions.[26]

Another dimension of the world of high finance in Malwa was the personal stake that Baiza Bai had in it in her capacity as a banker. It was common knowledge that Baiza Bai herself was involved in large-scale banking transactions. The huge profits that she earned from moneylending, discounting bills of exchange, speculation and various other financial/banking activities made her fabulously wealthy. Baiza Bai had a large network of banking firms that operated under various names. At Ujjain, for instance, she reportedly presided over two banking firms, Nathji Kishan Das and Nathji Bhagwan Das.[27] Baiza Bai also had a banking concern at Banaras. When the premises of the bank were temporarily sealed by the British following the coup against her, she protested that 'the inconvenience will be excessive, the Shurraffing or Banking concerns of my House will be ruined. The name of the firm will be lost.'[28]

Sindia's ruling and commercial elite were loathe to provide any details of their transactions to company officials. In a long note on Malwa sahukars penned by John Malcolm, he begins by commenting upon 'the jealousy with which the Native Governments and above all that of Dowlut Row Scindiah regard every interference that can affect those money dealings of its Soukars and Bunniahs which are connected with the most corrupt of all the sources of emolument enjoyed by the Ruler, his Ministers and local officers [i.e. speculation]'. Even an enquiry into something seemingly as innocuous as the woollen trade in Malwa was looked upon with extreme suspicion making it difficult to gather commercial information since, according to Malcolm, 'it excited an uneasiness and alarm among men who, cannot yet understand the motives which lead to a minute investigation on such subjects'.[29]

In her financial career, the hold of Baiza Bai over the city of Ujjain and the possession of personal landed estates (*khasgi*) in the Ujjain pargana brought her into the opium business.[30] Throughout our period, Ujjain remained the financial and commercial centre of Malwa while the Ujjain countryside had some of the most productive opium fields in the entire region. While the average yield of opium juice in Malwa around 1820, according to Dangerfield, was 5 *pakka ser* per bigha, the average yield in central and western Malwa, including Ujjain, Mandsaur and Ratlam, was more than double at 10–12 *pakka ser* per bigha.[31] Malcolm observed in 1821 that Ujjain was 'the chief commercial emporium of this country [Malwa]' and had been 'for a century the residence of the Agents of the Merchants in Hindustan, and the Deckan and Guzerat, who carry on the great transit commerce through Malwa'.[32]

In addition Ujjain was the foremost wholesale market for opium. Restrictions placed by the company did divert some of the trade away from the city during the 1820s but the larger financial operations related to the commodity were still concentrated there. Opium smuggling also encouraged speculative ventures in which significant sums of money were tied up. The interlinking of speculative ventures involving bills of exchange, grain, cotton and opium resulted in large-scale financial transactions. One such transaction involved cotton valued at between Rs. 1,080,000 and Rs. 1,278,000.[33]

A proportion of the available capital was channeled into high-risk speculation, which sometimes bordered on gambling, in opium stocks. Malcolm was convinced that 'the consequence of the gambling traffic carried on in opium have hitherto and will unless prevented continue to defeat the object which the British Government have in view of purchasing at a reasonable price the surplus produce of the Drug in

Malwa'. He added 'its present high price is entirely kept up by the Soukars of Oujein'.[34] A nominee of Gokul Parakh and a relative of the Sindia family carried out the actual administration of the city jointly. In one example of the close inter-linkages between the Sindia elite, Man Singh Patankar, whose son was married to one of Daulat Rao Sindia's daughters, was the *amildar* of the city for some time in the 1820s. It is no coincidence that he was engaged in opium smuggling as well.[35]

Although Baiza Bai was deposed in 1833 and exiled from Sindia territories, her banking concern continued to flourish. In 1840, she settled down in Nasik, where she stayed for 8 years before being allowed to return home. She had among her associates at Nasik a Bombay businessman named Bapu Joshi. Joshi had been a shipowner who had shifted to Nasik with the decline of the indigenous shipping industry in Bombay.[36] Baiza Bai's financial dealings might have extended to Bombay during her Nasik sojourn with Joshi acting as her agent. It was Bapu Joshi who successfully negotiated the end of Baiza Bai's exile in 1848. According to the provisions of the agreement drawn up on this occasion, she was once again given charge of Ujjain pargana, including the city.[37]

The available evidence suggests that by the 1830s the landed magnates of the Sindia territories had lost interest in opium. Unlike the situation that prevailed till the end of the 1820s, there are hardly any references to the direct involvement of high officials, big landowners or bankers in the opium enterprise in colonial records of the late 1830s. It seems that the speculative ventures that the clandestine trade had given rise to attracted massive investments in the drug. With the introduction of the pass system and the emergence of Bombay as the main outlet for Malwa opium, the returns were not sufficiently alluring for the Sindia elite to associate with the trade in a big way. For those investors who had an imperfect knowledge of the China market, dealing in opium was a high-risk venture. Otherwise too, the trade was subject to considerable fluctuations during the 1820s and the 1830s. Trocki has pointed out that 'the system lurched out of control' in this period and that 'there were a series of booms and busts in the trade and a constant tendency to speculate, which in the short run, made the trade seem very risky'.[38] Thus, the Maratha aristocracy of Malwa, a section of which had actively participated in the opium enterprise, did not become a major component of the modern business class of western India. Instead, having initially accumulated their capital by speculating in opium, the class of revenue farmers-cum-bankers-cum-officials formerly typified by people like Gokul Parakh at Gwalior and Tatya Jog (chief minister of the Holkar state) at Indore[39] preferred, by the mid-1830s, to make use of positions at the darbar to augment their incomes in less risky ways through revenue farming, interest on loans to the state, discounting bills of exchange and profits from military contracts. Not all profited and evidence of possible bankruptcy is the disappearance of Appa Gangadhar from the records in the 1830s. Such bankruptcies would have heightened fears about the financial dangers of the opium trade and possibly explain the switch to revenue farming in the 1830s.

A new breed of bankers began to dominate the Sindia economy from the 1830s onwards, one of the most outstanding of these being Mani Ram. He had been active in the decades preceding the 1830s and had close business connections with Parakh.[40] He had built a vast financial empire under Baiza Bai's regency and had emerged as the foremost banker in the Sindia state and perhaps in the whole of Malwa. The Company often had to rely on him for credit for its financial requirements in Malwa.[41] Mani Ram was the treasurer for the tribute collected by the Sindias from states in Rajasthan. He also held the revenue farm for the entire Gwalior suba and by 1833 the outstanding dues for this particular *ijara* amounted to about 10 million rupees. Baiza Bai's insistence that he pay up his dues was one of the causes of the coup against her.[42] The stability that this class had acquired, a development in which earlier earnings from opium played no insignificant part, made it look towards more secure returns from long-term linkages with the state. It also meant that this class had a stake in maintaining the autonomy of the Sindia state, an interest that lead eventually to a breakdown in relations with the company and the military campaign of 1843–1844.

Opium along the west coast

Ironically it was the smuggling trade itself that speeded up the shift away from Ujjain since the major smuggling routes passed through north-western Malwa.[43] The distance of the city from the remote northern Malwa parganas would have led to smaller quantities of the drug being warehoused in it. Moreover, the physical proximity of Ujjain to neighbouring Indore, where the company maintained its main civil establishment in Malwa, rendered the place unsafe from the point of view of opium transactions. Not only were the happenings at Ujjain closely scrutinised by colonial officials but there were actual attempts to undermine the commercial and financial status of the city.[44] The lack of strong support from the darbar in the post-Baiza Bai period sealed the fate of Ujjain.

Pali in Rajasthan, where the smuggling routes now converged, soon became the main centre of the wholesale trade in the western Indian hinterland. Pali was essentially a transit point where the earnings from opium were in the form of commissions and brokerage. Yet the shift to Pali provided Rajasthani merchants with crucial advantages through which they could establish themselves as a key component of the entire network: otherwise they might have remained marginal to it. These are the merchants of Rajasthan proper and not the traders of Rajasthani origin, including the Marwaris, settled in Malwa for many generations. However, the family and caste connections of the Pali dealers with Rajasthani merchants resident in Malwa were vital for the expansion of their businesses. The Gosains, who as armed ascetics had considerable social and religious influence within Ujjain, had also established a powerful base in southern Rajasthan where they dominated the transportation of opium from Malwa to Gujarat. They maintained a retinue of Gosain informers and often relied on the support of the Charuns to carry opium to Pali.[45]

One group that was drawn into the ever-widening scope of the opium-trading network solely due to rampant smuggling was that of Sindhi merchants. Sind was located at too great a distance from Malwa to have otherwise become a part of this network. There were no traditional links between the spheres of activity of business groups from the two regions. Recent research has shown that a unique set of circumstances brought Sind within the orbit of the international opium economy. For nearly a quarter of a century, Karachi in Sind was the main entrepot in the Arabian Sea which handled short-distance shipment of the drug from producing areas in Malwa to west coast ports like Daman for the onward voyage to Southeast Asia. Some of the factors that contributed to this development were (i) the proximity of Sind to Rajasthan on the one hand and to the Gujarat coast on the other; (ii) the favourable location of Karachi port; (iii) the presence in Lower Sind of an enterprising class of merchants with extensive financial and commercial connections; (iv) the support extended to this class by the Talpur Amirs who ruled over Lower Sind; and (v) the lateness of colonial penetration of the region. Some scholars have gone so far as to suggest that the primary cause for the annexation of Sind by the company in 1843 was the continued leakage of Malwa opium through Karachi during the 1830s after the introduction of the pass system.[46] In 1840, soon after the British took over Karachi (the town was occupied in 1839), the company's senior officials in Calcutta were already contemplating an increase in the duty for opium passes as they were convinced 'that advantage in this respect may be obtained from the position which we at present occupy in Sind'.[47]

The Talpur Amirs carefully nurtured this trade, thereby augmenting the meagre resources to which their state had access. The customs revenue collected at Karachi in 1809 amounted to Rs. 99,000.[48] In 1823 the revenue farm for the port was contracted for Rs. 260,000.[49] It is likely that by 1830, opium duties alone were yielding Rs. 540,000 annually.[50] In 1837 the total customs revenue was Rs. 173,893, of which Rs. 65,000 came from opium. The transit trade had declined by this time but was still large enough. On an average 500 camel loads of opium, equivalent to approximately 1000 chests, passed through Karachi every year in the latter half of the decade. This was one-third of the average figure prior to the pass system.[51] In the words of Markovits, 'opium revenues were a major contribution to Sind finances'.[52]

As for the Sind merchants, their earnings came chiefly from the profits that they earned by arranging for the transportation of opium from the edge of the Thar desert, taking over from the Rajasthani middlemen to Karachi and from commissions on the remittance of opium duties to the state.[53] The Karachi traders also used their fleet of sturdy seafaring boats, the *dhangis*,[54] to ship the opium to Daman and other Gujarat ports.[55] An 1839 list of boats belonging to Karachi harbour mentions 28 boats, of which 27 are classified as *dhangis*. These boats had 14 different owners.[56] The owners included Seth Naomull, the leading merchant of Lower Sind, who was the largest owner with five boats.[57] Another owner, Ali Reekha Jemedar (Jemadar Alla Rukha), had three boats[58] and had been active in shipping opium from Karachi to Daman throughout the 1820s. He is first mentioned in a letter of 1822 at which time he was the *thanedar* of Karachi.[59] An informer of the company filed a report from Karachi in that year stating that one of Alla Rukha's *dhangis* was about to sail with a cargo of opium. The said opium was the property of two merchants, one from Surat and the other from Bombay.[60] The big profits were, of course, going to Bombay and Gujarat merchants but, as long as opium smuggling thrived, Sindhi merchants could have a small share of them.[61]

One other port on the west coast of India that attracted a large traffic as a consequence of opium smuggling was Portuguese Daman. Whereas the remoteness of Karachi gave to its business groups a degree of independence, Daman was actually located on the Gujarat coast and was physically close to Bombay. This made it difficult for the Portuguese enclave to escape the direct influence of the Gujarat and Bombay merchants. Beyond a certain point these merchants were not dependent on the collaboration of Indo-Portuguese groups for managing the unloading and loading of opium at Daman or its shipment to China. They could do this through their own agents. In other words the local connections of Indo-Portuguese traders did not bring to them any special advantages. Whereas the support of the Portuguese at Macao was vital, this might not have been the case at Daman. Here the local Gujarati network, operating in close association with indigenous and private European interests at Bombay, was powerful enough to seize upon the space provided by a Portuguese possession. To what extent the Indo-Portuguese traders comprised a distinct group at Daman is not certain. But it is doubtful that this was a very cohesive group since quite a few Indo-Portuguese operating out of Daman, such as Roger de Faria and Antonio Pereira, had their roots in Goa. The Gujarati banias and Parsis were better placed to convert the port into a base for their activities. Nevertheless the political support of the Indo-Portuguese was useful in keeping the trade at Daman legal, although in this they were pursuing their self-interest. As for Macao, already by the early eighteenth century, Portuguese traders here had their well-defined commercial schemes which could be at variance with the operations of the Estado da Índia on the west coast of India.[62] Daman never graduated to the status of a major port, neither did the Daman Indo-Portuguese traders as a group acquire stable wealth from their opium transactions. The successful career of Roger de Faria, the most well-known of the Indo-Portuguese opium merchants, ended in bankruptcy in 1838. His bankruptcy was the outcome of his failure to establish a firm control over the administration of Daman. That he should have taken the risk of supporting the government of Bernardo Peres da Silva, which had been overthrown at Goa and was forced to seek shelter in Daman, points towards a desperate attempt to recover lost ground in the 1830s.[63] It is significant that in return for the financial assistance extended by de Faria to a plan for the restoration of da Silva he was to receive exemptions on customs duties. These in turn were to be passed on to Bombay merchants apparently in the hope that de Faria's firm would continue to attract a share of the opium business.[64] In the long run, however, the Indo-Portuguese traders were too weak to benefit much from the short-lived importance of the Daman route.

Bombay merchants

The diversion of a large proportion of Malwa opium to Bombay through the pass system was not an unqualified success for the company. More than the company it was the Bombay business class that benefited from this move. The pass system in effect reduced the share of other indigenous groups and correspondingly increased the profit margins of Bombay dealers. For almost two decades, Bombay merchants had a virtual monopoly over the supply of the Malwa drug to China. This only began to change during the 1850s when there was a steep rise in the opium duty. The business class of Bombay cannot therefore be regarded as collaborationist in any real sense. No wonder the colonial state remained so hostile to it. It constantly tried to ensure that Bombay exporters received no support whatsoever from the government.

An incident that is illustrative of the company's grudging acceptance of competition from Bombay merchants occurred against the backdrop of the crisis in the Bengal opium trade on the eve of the First Opium War. In 1837, Bengal opium faced a crisis due to the difficulty that Calcutta buyers had in realising adequate profits on the sale of the commodity in China.[65] In this situation the government decided to refund to purchasers of Patna and Banaras opium a part of the auction money. The refund was, on an average, Rs. 200 per chest.[66] The Bombay Chamber of Commerce protested that 'in their opinion a departure by a seller from the published condition of sale is a violation of these conditions and though the individual benefited may not consider it so, other interested parties and the public generally are justly entitled to regard it as a breach of faith towards them'.[67] In a sharply worded rejoinder the Calcutta authorities hinted that Bombay merchants were actually interlopers and deserved no sympathy. What the company resented most was the extraordinarily large share that Bombay exporters pocketed, which amounted to depriving the government of a substantial portion of its potential revenue:

> The Merchants at Bombay cannot, it is presumed, be ignorant that, of the price paid by purchasers at the opium sales of Calcutta, about 300 Rs is the cost price of each chest of opium sold and the remainder is the Govt. duty. As the first sales of the past season therefore ranged at prices exceeding 1550 Rs for Behar opium the Govt. duty upon each chest of this opium exported was 1250 Rs or ten times that levied by the Government of Bombay upon each chest of Malwa opium exported through that Presidency.[68]

The supreme government's letter added that it could not be expected to look on helplessly 'while a trade which employed half the capital of the place and yielded to the Govt. a revenue exceeding in the present year two crores [20 million] of Rupees was ruined by the competition of dealers in a foreign article produced at twice the cost and taxed only with one tenth of the duty'.[69] The comparative economics of Bengal and Malwa opium wherein the former yielded to the colonial state almost 10 times the revenue per chest over the latter is indeed revealing and goes a long way towards explaining the urgency that the 'opening of China' had acquired by the end of the 1830s. With local merchants retaining most of the spoils of the opium trade, the British were limited to what they could earn from the imposition of licenses on opium bound for Bombay. This was a major embarrassment to British officials, clearly countermanding a major imperial imperative of controlling trade for the benefit of the colonisers. Thus, the British annexation of Sind and the First Opium War may then be seen as an attempt to resolve the contradictions arising from colonial intervention in the economy of western India whereby local groups were making larger profits from the opium trade than the colonial administration.

Notes

1. 'Net Proceeds of the Opium Sales in 1822', 29 August 1822, National Archives of India (hereafter NAI), Separate Revenue Branch Consultations (hereafter SRBC), 26/19 December 1822; Minute by F. Warden, Secretary, Bombay Government, 30 April 1823, 56/12 June 1823; J.F. Richards, 'The Opium Industry in British India' in *The Indian Economic and Social History Review*, XXXIX (April–September 2002) Table 1.1, p. 160.
2. Richards, 'Opium Industry', p. 174.
3. Report of F. Dangerfield, enclosed in letter to John Malcolm, 3 October 1820, NAI, SRBC, 18/1 December 1820 (henceforth Dangerfield's Report).
4. For more details, see Amar Farooqui, 'From Baiza Bai To Lakshmi Bai: The Sindia State in the Early Nineteenth Century and the Roots of 1857' in B. Pati (ed.), *Issues in Modern Indian History: For Sumit Sarkar* (Mumbai: 2000), pp. 45–74.
5. The evolution of the residency system is discussed in M.H. Fisher, *Indirect Rule in India: Residents and the Residency System, 1764–1858* (Delhi: 1991).
6. Cf. R. Cavendish, Resident, Gwalior, to W.H. Macnaghten, Secretary, Govt. of India, 14 July 1832, NAI, Foreign Department (hereafter FD), Misc., Vol. 235.
7. Cavendish to Macnaghten, 29 May 1833, NAI, FD, Political, 42/13 June 1833; J. Stewart, Resident, Gwalior, to Govt. of India, 23 March 1830, NAI, FD, Political, 54/16 April 1830.
8. J. Malcolm, *A Memoir of Central India, Including Malwa and Adjoining Provinces*, Vol. I (London: 1823; reprinted New Delhi, 1970).
9. See A. Farooqui, *Smuggling as Subversion: Colonialism, Indian Merchants and the Politics of Opium* (Delhi: 1998), pp. 101–2.
10. For example, Baiza Bai avoided formally receiving a letter addressed to her by the Governor-General whereby the company tried to interfere in court

ceremonial. The Governor-General had outlined measures that would have undermined her position vis-à-vis her adopted son, for whom she was the regent. G. Fielding, Acting Resident, Gwalior, to Govt. of India, 18 January 1830, NAI, FD, Political, 45/16 April 1830.

11. Lt. Col. W.H. Sleeman, Agent to the Governor-General, Sindia's Dominions, to Govt. of India, 21 October 1843, NAI, FD, Political, 220/23 March 1844.
12. Unfortunately the available documents of the Sindia state in the Gwalior Record Room of the Madhya Pradesh State Archives, Bhopal, do not provide figures which could be used for the purpose of estimating the revenue from opium. There are no centralised statistics for this period and land records are mostly title deeds.
13. 'Statement of the number of Beeghas of opium cultivated in the Pergunnahs in Malwa', 15 November 1823, NAI, SRBC, 16/5 December 1823; S. Swinton, Opium Agent, to Board of Customs, Salt and Opium (hereafter BCSO), 15 February 1826, NAI, SRBC, 45/9 March 1826; Swinton to BCSO, 25 May 1826, NAI, SRBC, 21/22 June 1826.
14. W.E. Cheong, *Mandarins and Merchants: Jardine Matheson & Co., a China agency of the early nineteeth century* (London: 1979), p. 21, Table III; Dangerfield's Report; C.E. Luard, *The Central India State Gazetteer Series, Vol. V, Western States (Malwa) Gazetteer, Part A* (Bombay: 1908).
15. Dangerfield's Report.
16. Methods of poppy cultivation observed by Dangerfield around 1820 are identical to those described in the, compiled at the beginning of the twentieth century, *Western States (Malwa) Gazetteer, Part A*, p. 19.
17. Swinton to BCSO, 26 May 1826, NAI, SRBC, 23/22 June 1826.
18. *Khunt* developed into an impost on property for protection granted within the limits of a certain area.
19. Dangerfield's Report.
20. Cf. Farooqui, *Smuggling as Subversion*, p. 113.
21. Letter of Sleeman, 21 October 1843, NAI, FD, Political, 220/23 March 1844.
22. S.C. Macpherson, Political Agent, Gwalior, to R. Hamilton, Agent, Central India, 13 December 1856, NAI, Central India Agency Records, Gwalior Residency, no. 267, pp. 4–5.
23. Cf. Farooqui, *Smuggling as Subversion*, pp. 45–48.
24. 'Paruk and the late Maharaja are said to have been partners in the bank'. Cavendish to Macnaghten, 7 February 1834, NAI, FD, Political, 13/3 April 1834.
25. J. Sarkar (ed.), *Poona Residency Correspondence*, Vol. XIV (1810–1818), p. 297, letter no. 251; 'Note on Malwa sahukars', John Malcolm to Capt. R. Close, Resident, Gwalior, NAI, SRBC, 2/23 April 1821.
26. Swinton to BCSO, 24 April 1826, NAI, SRBC, 11/22 June 1826.
27. Capt. O. Stubbs, Sindia's Contingent, to Maj. G. Fielding, Acting Resident, Gwalior, 20 February 1829, NAI, FD, Political, 50/15 May 1829.
28. Baiza Bai to Governor-General, tr., n.d., NAI, FDP, 18/19 October 1835.
29. Malcolm to Bombay Govt., 15 November 1819, NAI, FD, Political, 36/18 December 1819.
30. Farooqui, 'Baiza Bai', p. 51.
31. Dangerfield's Report.
32. Malcolm to Close, 3 March 1821, NAI, SRBC, 2/23 April 1821.

33. Ibid.
34. Ibid.
35. Swinton to Trotter, 16 May 1826, NAI, SRBC, 19/22 June 1826.
36. R. Shakespear, Assistant for Sindia's Dominions, to Sleeman, 12 March 1848, NAI, FD, Political, 114/7 April 1848.
37. 'Agreement between the Baiza Bai and the Gwalior Darbar', tr., NAI, FD, Political, 121/7 April 1848.
38. C.A. Trocki, *Opium, Empire and the Global Political Economy: A Study of the Asian Opium Trade, 1750–1950* (London: 1999), p. 94.
39. On the eve of his departure from Malwa, Malcolm commented that Tatya Jog, who had been 'brought up as a sahokar and [was] subsequently raised to power as a Minister in the very worst of times', was 'from habit fond of making money'. Malcolm to Govt. of India, 5 August 1821, NAI, FD, Political, 23/15 September 1821.
40. Cf. Malcolm to Close, 2 March 1821, NAI, SRBC, 2/23 April 1821.
41. Dixon Dyke, Acting Resident, Gwalior, to Govt. of India, 29 October 1830, NAI, FD, Political, 21/7 December 1830.
42. Cavendish to Macnaghten, 23 May 1833, NAI, FD, Political, 37/13 June 1833.
43. See Farooqui, *Smuggling as Subversion*, pp. 118–20.
44. Cf. Malcolm to Close, 3 March 1821, NAI, SRBC, 2/23 April 1821.
45. A letter intercepted by the company's officials sheds light on their well-knit organisation in Rajasthan and Gujarat. Letter from Gosain Goverdhinghur and others of Purtapghur to Gosains Rutughur Ramghur and others at Ahmedabad, tr., enclosed in John Dunlop, Collector, Ahmadabad to James Williams, Resident, Baroda, 5 June 1820, Maharashtra State Archives, Mumbai (hereafter MSAM), Revenue Department (hereafter RD), Vol. 9/9.
46. The boldest assertion of this argument is to be found in J.Y. Wong, 'British Annexation of Sind in 1843: An Economic Perspective' in *Modern Asian Studies*, XXXI, 1997, pp. 226 and 234–7. A more nuanced assessment is presented by Claude Markovits in his study of the Sind merchants: '[T]hat some correlation existed between British opium policy on the one hand and the decision to annex Sind seems indubitable, but it does not prove that the desire to close the Sind route to Malwa opium was the main motive of the annexation'. Markovits does, however, emphasise that 'the occupation of Karachi by British troops in 1839 allowed the East India Company effectively to close the Pali–Karachi route and to redirect the Malwa opium trade through Bombay, which they had tried, unsuccessfully, to do for almost twenty years'. C. Markovits, *The Global World of Indian Merchants, 1750–1947: Traders from Bukhara to Panama* (Cambridge: 2000) p. 41 and n. 23.
47. Govt. of India to Lt. Col. Sir C.M. Wade, Resident, Indore, 12 August 1840, NAI, SRBC, 13/19 August 1840. The British occupation of Karachi in 1839 was supposed to be a temporary measure, but the Talpur amirs 'were never destined to recover' the port. See R.A. Huttenback, *British Relations with Sind, 1799–1843: An Anatomy of Imperialism* (Berkeley: 1962), p. 54.
48. Selections from the Records of the Bombay Government, new series, no. XVII, in two parts (hereafter BGS, XVII) Bombay Govt., Bombay, 1855, part i, p. 198.
49. T.G. Gardiner, Resident, Cutch, to Bombay Govt, 22 January 1823, NAI, SRBC, 11/6 March 1823.

50. Letter of Henry Pottinger, Resident, Bhuj [Kachchh], cited in Markovits, *Global World of Indian Merchants*, p. 42.
51. BGS, XVII, part i, pp. 198 and 201.
52. Markovits, *Global World of Indian Merchants*, p. 42.
53. Ibid.
54. The *dhangi* was described as 'sharp bowed, bottom and stern'. It resembled the lateen-rigged *patimar* that was so common along the west coast 'excepting having a high stern and poop'. Its advantage lay in its speed. Cf. *BGS*, XVII, part i, p. 267.
55. At the same time ships were sent up from Daman, or Gujarat ports such as Mandvi, to bring opium from Karachi.
56. *BGS*, XVII, part i, p. 267.
57. Ibid. In the list the name of Naomull Hotechand appears as 'Naomull Moteram' which is probably the result of careless copying. On Seth Naomull see *BGS*, XVII, part ii, p. 667; Markovits, *Global World of Indian Merchants*, pp. 35–6.
58. Ibid.
59. What this position meant is uncertain, but it must have carried with it policing duties.
60. Ayoob Hasun at Kurachee Bunder to Ruttonsee Jaita, 13 January 1822, enclosed in Resident, Bhuj, to Bombay Govt., 19 March 1822, MSAM, RD, 18/42.
61. According to Markovits 'it is striking that the major trading houses of Sind *banias* in Karachi, such as those of Seth Naomal Hotchand..., went into rapid decline after the annexation [of Sind]'. Markovits, *Global World of Indian Merchants*, p. 55.
62. I have been led to revise my understanding of the nature of Indo-Portuguese involvement in the opium trade at Daman. The picture is much more complex than I had outlined earlier (*Smuggling as Subversion*, pp. 122–127). For a discussion on the contradictions between Macao and Goa, in relation to Portuguese trade at Surat, see G.B. Souza, *The Survival of Empire: Portuguese Trade and Society in China and the South China Sea, 1630–1754* (Cambridge: 1986), pp. 177–81.
63. Bernardo Peres da Silva was the first indigenous Goan appointed to head the Estado da Índia, and his official title was 'Prefect'. He assumed office in January 1835, but was ousted within seventeen days following a revolt in Goa. F.C. Danvers, *The Portuguese in India*, 2 vols. (London: 1966; first edition, 1894), Vol. II, p. 457.
64. For details, see T.R. de Souza, 'Rogerio de Faria: An Indo-Portuguese Trader with China Links', A.T. de Matos and L.F.F.R. Thomaz (ed.), *As Relacões entre a Índia Portuguesa, a Ásia do Sueste, e o Extremo Oriente* (Macau and Lisbon: 1993), pp. 309–19.
65. On the causes of this crisis, see Trocki, *Opium, Empire and the Global Political Economy*, p. 94.
66. Minute of Lord Auckland, Governor-General, 29 July 1837, NAI, SRBC, 6/9 August 1837; Bombay Chamber of Commerce to Govt. of India, 3 September 1837, NAI, SRBC, 4/27 September 1837.
67. Bombay Chamber of Commerce to Govt of India, 3 September 1837, NAI, SRBC, 4/27 September 1837; Bombay Chamber of Commerce to Govt. of

India, 6 November 1837, NAI, SRBC, 2/22 November 1837. The refund was a matter of concern for the Bombay merchants because prices at Calcutta determined the prices that they paid for Malwa opium.
68. Govt. of India to Bombay Govt., 27 September 1837, NAI, SRBC, 6/27 September 1837.
69. Ibid.

7
Dangerous Drinks and the Colonial State: 'Illicit' Gin Prohibition and Control in Colonial Nigeria

Chima J. Korieh

Introduction

> The public in general greatly resents the prohibition of home-made distilled liquors, and look on it as an unwarranted interference with their legitimate activities (District Officer, Degema, Owerri Province, 30 December 1938).[1]

This chapter will examine the reasons that the British colonial administration in Nigeria criminalised locally produced spirits in the 1930s and 1940s, with a particular focus on locally produced gin (*ogogoro, kai kai, Sapele water, agbagba, akpeteshi, aka mere, push me I push you, and crazy man in the bottle*) as it is commonly called in southern Nigeria. Under the Sections 8 and 9 of Cap. 131 (the Liquor Ordinance of 1931), anyone convicted of producing, distributing or possessing local gin was liable to a fine of 100 pounds or 6 months imprisonment, and under Section 60 of the Ordinance, a person who had been previously convicted of a similar offence could be sentenced to 2 years imprisonment. This mechanism was central to what became an aggressive policy to control local production of spirits that was pursued from 1931 to the mid-1940s. However, the chapter will also examine local resistance to these policies. It will argue that the administration's stance only succeeded in producing a situation in which alcohol could become a site for the defiance of the colonial regime.

Scholars have paid attention to convoluted politics of alcohol in colonial Africa[2] but relatively little has been written about state control of local liquor production during the colonial period. While significant attention has been given to the role of alcohol in empire, the omission of the effort to control what colonial officials termed 'illicit' distillation is surprising given the importance the British attached to limiting local production. Exceptions include Emmanuel Akyeampong's detailed account of the attempt by the British colonial authority to control the distillation of spirits in Ghana during the 1930s, the response of the local people and its link to the anti-colonial sentiment of the post-Second World War era.[3] Ayodeji Olukoju has also examined British policy towards liquor traffic in Moslem Northern Nigeria.[4] The analysis here not only focuses on a largely unexplored aspect of rural struggle in colonial Nigeria but also sheds light on the important link between colonial revenue and the consistent effort to characterise local distilled spirit as a dangerous development and threat to the health of the local people.

Alcohol and Nigeria

Across much of what is now, Nigeria people drank (and still drink) palm wine, usually sourced by professional wine tappers, and beer made from malted grains. Indeed prior to the colonial period, alcohol was widely used and enjoyed throughout Nigerian societies. The drinking of different types of alcoholic beverage was a socially accepted practice which was often embedded within the social fabric of local societies; drinks were used to build and reinforce relationships and kinship networks and were also a feature of formal cultural or ritual occasions. Suitors offered liquor to the kin of their prospective brides, men sent gifts of alcohol to their in-laws and subjects would present drink to their chiefs. Alcohol was poured in libations to ancestors when the living asked for help in the face of disease or misfortune; it was given to and drunk by the elder men who played a central part in ritual events; 'given drink, men might use it to bless their juniors by taking it in their mouths and "blowing" it over them, thus expressing their renunciation of any anger'.[5] In East Africa, alcohol was a symbol of authority and a source of status and privilege in local societies.[6] Due to expanding European activities in Western Africa, imported alcoholic drinks entered these drinking traditions and were also an important part of the new consumer culture that began to emerge there from the fifteenth century. European contact brought in a variety of alcoholic beverages, primarily gin, brandy, whisky and beer. Their introduction altered the tastes and consumption patterns but also opened up a competition between various locally produced alcoholic beverages and the imported brands.

European traders on the West Coast of Africa exchanged alcohol for local goods and slaves from the fifteenth century onwards. As the

European presence in Africa expanded, European producers and trading houses exported large quantities of alcoholic beverages and merchants in Liverpool set up distilleries specifically to cater for the African trade. The end of the slave trade and the emergence of the commodity trade significantly increased the importance of alcohol in the local economy.[7] At the Nigerian coastal trading posts, alcohol became so highly valued as a means of exchange that a Foreign Office official there wrote in 1895 that without spirits 'trade in the Delta is at present impossible'.[8] German commercial traders were exporting about 45 million litres of spirit to sub-Saharan Africa in the middle of the nineteenth century.[9] Spirits remained the most important German export to West Africa until the colonial period.[10]

At the social level, imported drink, especially Schnapps, became part of the local tradition. *Mmanyi Oku* (hot drinks), as they are popularly known among the Igbo of Nigeria, became a requirement in marriage ceremonies, religious rituals and an important gift item to elders and patrons. The colonial encounter expanded the use of alcohol. The British used alcohol to build alliances with the traditional elite. In the Cameroon, the Germans used alcohol to pay for labour.[11] Unlike many other commodities, therefore, alcohol reinforced the link between the state and the local polities.

Yet while alcohol was both a valuable commodity for European traders and a tool of imperial policy, it assumed even greater significance by the end of the nineteenth century, as a substantial portion of official revenue was derived from customs duties and tariffs imposed on alcohol.[12] Olorunfemi has noted that over half of the total revenue in Nigeria after the 1880s came from duties and tariffs imposed upon imported liquor.[13] Similarly, Heap has highlighted the importance of liquor revenue to the colonial state in Nigeria.[14] The European authorities came to perceive any threat to their liquor revenues as a threat to the colonial state itself.

Alcohol imports and the colonial state

Since the beginning of the colonial period, governments had made laws to control the manufacture, sale and consumption of alcohol in many parts of Africa. Their policies, however, were often drawn in different directions because of the competing forces that underpinned them. On the one hand, there was an impulse to deny alcohol to Africans. This can partly be explained by state anxiety about disorder which in Africa grew out of a concern that drunken natives would defy the colonial state or flout the authority of their own elders and chiefs. In East Africa, for example, British colonial authorities forbade Africans to drink bottled beers and wines, and also banned spirits of all kinds, believing that European beers and wines were more potent than traditional liquor.[15]

The impulse to deny alcohol to Africans also rested upon the discourses of the 'civilizing mission'. The concern with alcohol can be seen in the Brussels General Act of 1889–1890 which, while primarily concerned with measures against the slave trade, included a chapter on 'Restrictive Measures Concerning the Traffic in Spirituous Liquor'.[16] It can also be seen in the activities of humanitarian organisations such as the Aborigines Protection Society. They denounced the liquor traffic in the same way that they denounced the slave trade,[17] arguing that the liquor shipped to West Africa 'contained poisonous alcohol injurious to health and widespread drunkenness and laziness'.[18] Indeed the Aborigines Protection Society contended that the alcohol trade 'gave rise to wasteful expenditure, immorality and crime', and thus to 'wholesale degradation of tribes and communities'.[19] In their view, the importation of alcohol in the territories contradicted the European civilising mission in Africa. Such humanitarian groups also had powerful economic allies; the anti-liquor lobby was led by the cotton textile merchants represented by the Manchester Chamber of Commerce who wished to increase their share of the markets in colonial Africa.

While there were those who thought that colonial laws should be employed to stop Africans engaging in dangerous consumption, others saw the duties of those laws as simply to protect the revenues of the groups that stood to profit from selling liquor to colonial markets. As already mentioned, many colonial governments saw revenue possibilities in controlling alcohol markets; the most extreme examples of this come from Nairobi (from 1922) and Mombasa (from 1934) where the authorities operated beer hall monopoly systems in which they were the sole providers of liquor to the African population. There was also a strong 'liquor merchant' lobby who regarded the liquor trade as an extension of the 'legitimate commerce' that followed the abolition of the Atlantic slave trade.[20]

In what became Nigeria, European policy on alcohol was very ambivalent from the time of the Royal Niger Company in the late nineteenth century onwards. Simon Heap has documented the various attempts to stamp out sales in northern Nigeria because Muslim beliefs prohibit the use of intoxicating liquor. In order to maintain good relations with the Sokoto Sultan in the north, the British under Fredrick Lugard established a policy of religious non-interference towards the Muslims in

1903. A raft of measures to ensure this included a ban on Christian missions to Muslim areas and a duty of 4 shillings per gallon on alcoholic beverages in the inland regions north of the 7th parallel (imposed as early as July 1890). Colonial policy for the north therefore focused on the vigorous control of alcoholic drinks to protect the area from the negative effects of drink; the policy was motivated by a colonial political desire to mollify a regional power.[21]

In the south of the territory where there were no such Islamic sensibilities to accommodate, both Christian missions and liquor were allowed free rein. The import of liquor to Nigeria grew as colonial control over the territory became more effective. Liquor was the most important import and provided the bulk of state revenue during the early colonial period. Between 1901 and 1910 the average amount of spirits imported annually was about 3 million gallons, mostly Dutch 'trade gin'.[22] In 1914, 'Nigeria imported over 4 million gallons of beverages, chiefly German and Dutch schnapps'.[23] Needless to say, not all of this alcohol was consumed by the markets of the south; as Heap has argued, the attempt to prohibit the sale of alcohol in northern Nigeria was thwarted by 'African ingenuity and deliberate efforts combined with colonial expediencies'.[24]

The British were little concerned about this leakage into the northern territories and alcohol regulations followed a mundane pattern until the 1930s. British passed legislation in 1920 governing the importation of Spirituous Liquors into the West African colonies (Spirituous Liquor Ordinance (Cap. 115)). The Ordinance was passed in order to give effect to the provisions of the Convention relating to the Liquor Traffic in Africa, which was signed at Saint Germain-en-Laye on 10 September 1919. Under the Ordinance the importation, sale and distribution of 'traffic spirits' and 'injurious spirits' were prohibited.[25] In 1923 the Spirituous Liquor Rules were passed under which spirits containing more than 65 per cent of absolute alcohol and trade spirits were not permitted to be imported, sold, disposed of or distributed in the colonies. In 1928 the Spirituous Liquor (Amendment) Rule No. 4 of 1928 permitted the Receiver General to accept Composite Certificates of the age and origin of brandy, rum or whisky. In the same year, the Spirituous Liquor (Amendment) (No. 2) Rule (No. 10) provided a new definition of 'Gin' and required in the case of pot-still gin a certificate of origin by the Competent Government Officer in the country of production. In 1932 the Spirituous Liquor Rule (No. 7 of 1932) provided that no brand of gin shall be notified as an approved brand unless the bottle in which the gin is contained was labelled with the name and address of the owner.[26]

However, during the early 1930s the worldwide depression and resulting financial crisis affected the economic fortunes of colonial Nigeria. The total value of imports and exports in 1931 amounted to a decrease of 'over ten million [pounds] from the figure for 1930'.[27] The depression resulted in low wages and a decline in the prices of local produce including foodstuffs, but without a corresponding fall in the price of imported items. The high cost of imported alcohol and the decreased volume of imports made it difficult for local users to maintain their habits during this period. In the Colony and Protectorate of Southern Nigeria, 548,128 gallons of gin was imported in 1927. By 1931, this had dropped to 331,833 gallons and this had fallen to 59,312 gallons by 1934. The total value of imports and exports in 1931 amounted to £17 million, a fall of over £10 million from the previous year.[28] The massive decline in the Nigerian importation of spirits acted as a catalyst for the local distillation industry. A gap in the market opened and distillation techniques were rapidly improvised; the result was a locally produced gin. Local people initially lauded *Ogogoro* for its medicinal qualities and it was prescribed for the relief of cold and other ailments. It became an intrinsic part of rural life in coastal regions as a stimulant during fishing trips on the oceans and rivers, and as a feature of social life and rituals.

The earliest local distillation can be traced in the colonial correspondence of the 1930s. In a letter to the District Officer for Udi in the eastern region of Nigeria on 7 December 1931, a magistrate in Enugu noted that the area had no 'trace of the distillation of spirits'.[29] In that same year, however, colleagues elsewhere produced very different assessments; 'the practice of making native gin from palm wine is a new one but is becoming common', the senior Resident for Onitsha Province wrote in a public notice in 1931.[30] In the Calabar Province, the Colonial Resident, G.H. Findlay, wrote that 'reports of "illicit" distillation were first made in April 1931 following the increased demand for sugar, copper piping and an usually high rate of thefts of feed pipes from launches and cars'.[31] The copper piping and feed pipes were important parts of the distiller's equipment. The process of distilling local gin is quite simple and requires palm wine or fermented fruit juice and sugar, two bottles and a metal worm. The palm wine is fermented and boiled in a metal drum. The drum is connected to a copper pipe which is passed through a hollow wooden trough filled with water. The water acts as condenser. The distilled liquor is collected at the other end of the copper pipe. The simple nature of the production process meant that the gin could be made anywhere with ease, and that manufacture was difficult to detect.

The sudden and widespread distillation of local gin had immediate implications for colonial rule. British officials and traders feared that increased local distillation would threaten its revenue and the interests of European merchants. On 15 March 1932, the Inspector-General of Police Mr C.W. Duncan wrote to the secretary of the Southern Provinces urging that 'all administrative officers be warned to use every effort to detect and punish offenders within their jurisdiction and so prevent the spread of this menace to the country's revenue'.[32] He was not alone in voicing such concerns. Charles R. Miler, a collector of Customs, complained that:

> There is no doubt that the production of illicit distilled liquor is on the increase in the Calabar Province and is, I regret to say, spreading to the Cameroons Province. It is becoming a serious menace to revenue as it is competing severely with imported gin.[33]

It was not only colonial administrators who framed the phenomenon in these terms. The Manager of the United African Company at Ikot Ekpene wrote to the District Officer, on 28 January 1932, complaining that the annual sales of liquor in the branch had dropped from £1300 in 1929 to £450 in 1931.[34]

Members of the colonial administration also worried that local alcohol distillation could impact on the health of the colonised population. The Senior Resident for Onitsha Province wrote to the District Officer for Enugu Division on 25 November 1931:

> I have to inform you that the manufacture of alcohol by the distilling of palm wine and sugar is being carried on to an increasing extent and is liable to become a serious menace to the native community...The evil effect of an unlimited supply of crude alcohol on the Native population is a matter which gives rise to grave anxiety.[35]

He was convinced that 'gin or rum made in this bush fashion is very harmful and may make those drinking it mad or blind'.[36] Some found it difficult to distinguish between the financial and the medical; 'the principal argument against illicit distillation, apart from the fiscal one, is the danger to health caused by it', wrote one colonial official.[37] However, within the correspondence on the issue there also lurks the fear identified earlier, of alcohol induced disorder threatening to undermine colonial rule. The Commissioner of Police in Calabar, for instance, reported that one of his constables was himself arrested in a neighbouring village 'raving and maniac drunk and it is reasonably conjectured that his state could only have been caused by the consumption of locally distilled spirit'.[38]

While the authorities fretted, there also seems to have been pressure to act coming from within the Nigerian elite. For example, S.T. Maolu wrote to the District Officer of Aba in 1932 stating that

> the distillers of spirit-liquor in southern Nigeria by the natives both men and women are very very great, which help to make the revenue of Nigeria to come down; and the distillers still increasing more and more in member, and they can distil more than 3000 gallons [of] drink per day; and the Nigeria government has not taken the strongest step they are ought to take on the matter, to stop these dreadful liquor which are very injurious and dangerous to the human health, which are stronger and more powerful than the English-Liquor.[39]

Mr Maolu, who described himself as a lover of Nigeria and the British government, shows the disquiet among the English-speaking African community at the apparent helplessness of the colonial administration to impose itself on the local liquor producers.

Repression and resistance

On 20 August 1931, the police raided the house of one Chief Alfred Njoku of Akabo in the Owerri Province of southeastern Nigeria and arrested three individuals. They claimed to be American subjects travelling around the country teaching people how to produce whisky, gin, rum and wine. They had suggested that for a fee of £25 they could teach the Chief how to distil liquor. Among those arrested was 35-year-old James Iso, who was a native of Calabar who had joined the crew of an American cargo ship in 1923. He went to New York and worked as a houseboy at different times to three American families. James was said to have visited the houses of traffickers and learnt the art of distilling liquor.[40] Like many others who engaged in the production of local gin, he had seen an opportunity to make quick money by selling his knowledge and expertise throughout southern Nigeria. James embodied the tradition of personal entrepreneurship rooted in the local culture which had increased tremendously due to the opportunities created by the depression in the colonial economy. However, such local initiative ran counter to the economic interests of European liquor traders as well as those of the colonial treasury.

It was to stifle such local initiative that the British decided to prohibit the distillation and distribution of 'illicit gin liquor' under the Illicit Distillation Ordinance of 1931. The prohibition was followed by extensive enforcement campaigns; 'there has been a large number of prosecutions and convictions in the Calabar Province' reported the Senior Resident for Onitsha Province in 1931.[41] On 10 November 1931, three men were convicted in the Provincial Court at Onitsha under the Liquor Ordinance (Cap. 131) 'of distilling spirits and sentence to a fine of fifty pounds or four months imprisonment, each of which was the maximum penalty the District Officer could inflict for a first offence'.[42] In the court of the Resident the maximum penalty for a first offence was 100 pounds and for a subsequent offence 500 pounds. Yet some were dissatisfied at these arrangements and felt them too lenient. The Secretary of the Southern Provinces, F.P. Lynch, wrote to the Chief Secretary to the government in Nigeria on 1 October 1931 to recommend imprisonment of up to 10 years, with whipping, in addition to fines for distillation-related offences.[43] Police informants were paid 10 shillings for information leading to the arrest and prosecution of illicit distillers.[44] Between 1932 and 1935 more than 1460 persons were convicted for illicit distillation in parts of the Southern Provinces.[45] Enforcement continued throughout the decade; between January and June 1939, for example, the police in Aba District prosecuted 40 persons for illicit distillation out of which 24 persons were convicted.[46]

Such repression remained ineffective. In extreme cases the law was openly defied. On 28 April 1932, for example, the chief and people of Ndom Ebom in the Uyo District of the Calabar Province confronted the police, who had arrested 37 locals for possessing apparatus and liquor. The show of strength of the villagers forced the release of the distillers.[47] Elsewhere, resistance was more surreptitious. The District Officer for Degema in the Rivers Province confessed in 1938 that 'there is no doubt that a great deal of illicit distillation still goes on in this Division, particularly in the Brass area. Practically any person who wishes to obtain illicit liquor can do so without much difficulty.'[48] Colleagues made similar observations, the district officer for Calabar noting that 'illicit distillation is widespread and for every case discovered by the police I estimate that 50 go undetected. I do not consider that present measures are stopping or will stop the distillation of liquor though they may stop its sale in the market'.[49] A decade later, the colonial authorities acknowledged the same problems; prohibitions, prosecution and imprisonment only drove the industry underground and forced the distillers to seek ever more remote locations to practise their art.[50] The failure of repression was not only noted within the ranks of the colonial government. The Eastern Regional Committee of the Christian Council of Nigeria wrote in 1944 regarding what it viewed as the alarming rate and rapid increase of the 'Drink Traffic in Nigeria'.[51] It noted that 'the distillation if illicit gin is more widespread than ever, and in cosmopolitan towns the numerous drinking bars are a menace to the physical and moral wellbeing of the people'.[52]

This flouting of the law eventually prompted some administrative officers to propose that local gin should be produced legally. The licensing of local distillation was one way that the government hoped to 'kill the illicit trade'.[53] Mr Barton, the Comptroller of Customs and Excise, suggested that the minimum output of legal stills should be about 3000 gallons per annum. Setting up such an operation would cost about £1000, which would act to limit the number of producers and make it easier for the government to control them.[54] The government did not take any immediate action. The production of local spirits continued and seemingly on a larger scale, especially in the Port Harcourt province. On 18 November 1947, for example, the police there raided a chain of local distilleries along the Imo River seizing and destroying about 250 gallons of illicit gin, 800–1000 gallons of fermented palm wine and 47 stills.[55] In 1947, the Resident for the Rivers Province reported that 'there were reports that gin distillation was going on an extensive scale in the Ahoda Division'.[56]

Many natives believed in the intrinsic right to distil and use alcohol. This view was supported by many within colonial circles who saw the futility of the government's crackdown on local distillation. The Resident for Rivers Province noted in 1947 that:

> The people of the creek areas have been accustomed for many years to obtaining and consuming quantities of spirits with no apparent ill effects. Now that supplies are unobtainable it is in my opinion inevitable that they should resort to measures which will produce what they require . . . the law was designed to protect the revenue, not the people's morals.[57]

At the meeting of the Rivers Provincial authority in June 1947, it was suggested that 'the question of legalising spirits distillation should again be considered by government'.[58] The suggestion was ignored, and for the rest of the colonial era local distillation remained a contested issue and arena for local protest.

Conclusion

The story told in this chapter goes to the heart of the relationship between colonialism and intoxicants, as it is a story shaped by competing interests and characterised by change and ambivalence. In the first place the story is framed by the economics of the British in Nigeria. The state wished to promote the interests of European companies eager to sell their alcoholic products in a region where there was a well-established market for such commodities that was based on the social customs of local society. It also wished to secure its own revenue from taxing such a lucrative trade. There was an imperative to sell drink to Africans. However, the politics of the region circumscribed this straightforward premise; maintaining stable relations with the Muslims of the north meant forbidding the same of intoxicating products there, at least at an official level.

The moral debates about the nature of imperialism, however, clouded British attitudes to the alcohol trade. Since the 1890s the missionary lobby had maintained that the sale of alcohol to African societies was in contradiction to the spirit of the 'civilizing mission' with which the whole colonial enterprise was legitimised. Yet much of this assertion was based on an Orientalist assumption that those societies needed protection from intoxicants. As with all such assumptions it was ill-founded. It reflected the representation of the African as a simple and unchanging entity, unable to adapt to the arrival of modernity as represented by liquor produced using industrial techniques. It also reflected the representation of the African as essentially lazy and indolent, who scarcely needed the effects of alcohol to enter a work-free stupor but who certainly would not benefit from the encouragement of drink to do so. It also reflected a typically colonial failure to take into account the particular nature of local societies. After all, as has been argued already, Nigerian society had an extensive experience of alcoholic beverages that stretched long into the period before the British arrived and it needed no protection from substances to which it was long-accustomed.

That the economic and the moral, in conflict before the 1930s, could fuse in that decade is a reflection of the changing nature of British colonialism in Nigeria. With the rise of a local industry to produce liquor to rival that on which the state's revenues depended, the authorities borrowed many of the discourses of the moral critics of alcohol consumption. The African, endangered by modern techniques and his own ignorance, was putting society in peril by producing dangerous drinks. The underlying rationale for this set of ideas, however, was not moral but medical. The basis for such an argument was of course economic, as the colonial government decided to seek a ban on local alcohol products in order to maintain its tax income. The irony is that when officials realised in the 1940s that the best way to maintain a revenue from alcoholic products was simply to make the locally produced substances legal and therefore exciseable, the spurious medical logic of the previous decade proved to be too well established in local mindsets to allow change.

Underlying all of these economic, political, moral and medical forces were the power relations of the colonial context. Much of the shape of the story comes down to a simple set of struggles between a local society and an alien government. Tastes may have changed and consumption patterns may have altered, but fundamentally Nigerians were used to their drink and carried on securing appropriate supplies regardless of the policies and interests of the foreign concerns. These supplies were often from abroad as the region entered global trading networks in the late nineteenth century, but became localised as entrepreneurial spirits in Nigerian society acted to meet the demand left unsatisfied by trading conditions caused by the Depression of the 1930s. The colonial government meanwhile saw the regulation of the alcohol trade not just as a means of securing revenue, but also as an opportunity to demonstrate the effectiveness of its regime and as an occasion for controlling and disciplining the colonised population. Ultimately, therefore, alcohol exposes the nature of the colonial encounter in Nigeria as the illicit gin story demonstrates how limited was the British state's ability to govern the local population, and how the commonplaces of consumption could be the root of sustained resistance to colonial designs.

Notes

1. National Archive Enugu (hereafter NAE) ABADIST 14/1/169, File no. 275 'Minute of a conference with the Residents, Eastern Provinces at Government Lodge, Enugu', 13 December 1948.
2. More recent scholarly works include D. Fahy Bryceson (ed.), *Alcohol in Africa: Mixing Business, Pleasure and Politics* (Portsmouth, NH: Heinemann 2002); J. Willis, *Potent Brews: A Social History of Alcohol in East Africa, 1850–1999* (London: The British Institute in Eastern Africa 2002).

3. E. Akyeampong, 'What's in a Drink? Class Struggle, Popular Culture and the Politics of Akpeteshie (Local Gin) in Ghana, 1930–67' in *Journal of African History*, 37, 1996, pp. 215–36.
4. A. Olukoju, 'Prohibition and Paternalism: The State and the Clandestine Liquor Traffic in Northern Nigeria, c. 1898–1918' in *The International Journal of African Historical Studies*, 24, 2, 1991, pp. 349–68. See also Simon Heap, ' "We Think Prohibition is a Farce": Drinking in the Alcohol-Prohibited Zone of Colonial Northern Nigeria' in *The International Journal of African Historical Studies*, 31, 1, 1998, pp. 23–51.
5. 'Alcohol in East Africa, 1850–1999', http://www.dur.ac.uk/History/web/past.htm. Accessed, 4 September 2003.
6. Willis, *Potent Brews*.
7. Alcohol seems to have played a more significant role during the late nineteenth and early parts of the twentieth century than it did during the slave trade era. An important reason perhaps was the widespread nature of the new commercial relationship which involved the majority of the local population as opposed to a select commercial and political elite during the earlier trade in slaves. For an overview of the commercial transition from slavery to commodity trade, see Robin Law, 'The Historiography of the Commercial Transition in Nineteenth Century West Africa' in T. Falola (ed.), *African Historiography: Essays in Honor of Jacon Ade Ajayi* (London: Harlow 1993). See also Lynn Pan, *Alcohol in Colonial Africa* (Uppsala: The Scandinavian Institute of African Studies 1975), p. 8.
8. A.C. Burns, *History of Nigeria* (London: George Allen and Unwin 1929), p. 169.
9. S. Diduk, 'European Alcohol, History, and the State in Cameroon' in *African Studies Review*, 36, 1, 1993, p. 2.
10. R. Oliver and G.N. Sanderson, *The Cambridge History of Africa, c. 1870–1905*, volume 6 (Cambridge: Cambridge University Press 1985), p. 130.
11. Diduk, 'European Alcohol', p. 1.
12. See for example, ibid., p. 4; A. Olurumfemi, 'The Liquor Traffic Dilemma in British West Africa: The Southern Nigeria Example, 1895–1918' in *International Journal of African Historical Studies*, 17, 2, 1984, pp. 229–41.
13. Olurumfemi, 'The Liquor Traffic', p. 236.
14. S. Heap, 'Living on the Proceeds of a Grog Shop: Liquor Revenue in Nigeria' in Fahy Bryceson, *Alcohol in Africa*, p. 140.
15. For example, see C. Ambler, 'Alcohol, Racial Segregation and Popular Politics in Northern Rhodesia' in *Journal of African History*, 31, 1990, pp. 295–313.
16. L. Pan, *Alcohol in Colonial Africa* (Helsinki: Finnish Foundation for Alcohol Studies 1975).
17. Olorunfemi, 'The Liquor Traffic', p. 229.
18. Ibid.
19. Ibid.
20. On the impact of the commercial transition for this region, see C.J. Korieh, 'The Nineteenth Century Commercial Transition in West Africa: The Case of the Biafra Hinterland' in *Canadian Journal of African Studies*, 34, 3, 2000, pp. 588–615; Law, 'The Historiography'.
21. See Heap, ' "We Think Prohibition is a Farce" ', pp. 23–51.

22. NAE RIVPROF 2/1/17, 'Record of Meeting at Government House to Discuss the Legalizing of Distillation of Spirits', 24 May 1943.
23. Heap, 'Living on the Proceeds', p. 140.
24. Ibid., p. 24.
25. 'Trade Sprits' were defined as spirits imported for sale to natives and not generally consumed by Europeans. They include mixtures and compounds made with such spirits. These include distilled liquor containing essential oil or chemical products, which are recognised as being injurious to health, such as thujona, star anisa, benzoic aldehyde, salicylic esters, hyssop, absinthe and similar substances unless such spirits have been denatured.
26. See UK National Archives, CO554/96/2, File no. 21446, 'Liquor Control'.
27. Editorial Notes, *Journal of the Royal African Society*, 31, 124, 1932, pp. 322–38.
28. Ibid.
29. NAE UDDIV 2/7/1, File no. EN 78/1931, the Station Magistrate, Enugu to District Officer Udi, 7 December 1931.
30. NAE UDDIV 2/7/1, File No. EN 78/19/31, 'Illicit Distillation of Spirits' W.E. Hunt, Senior Resident, Onitsha Province, 24 November 1931.
31. NAE RIVPROF 2/1/15, 'Memorandum on Illicit Distilling of Sprits in the Calabar Province', G.H. Findlay, 21 September 1931.
32. NAE AFDIST 5/1/2, 'Illicit Distillation of Spirits' Inspector General of Police, to the Secretary Southern Provinces, Enugu, 15 March 1932.
33. National Archives Calabar (hereafter NAC) 'Illicitly Distilled Liquor' Collector of Customs and Coordinating Officer, Eastern Preventive Service, Victoria (Charles R. Miler) to Comptroller of Customs, Lagos, 7 November 1932.
34. NAC CADIST 3/2/129, File no 11/31, 'Illicit Distillation,' District Officer, Ikot Ekpene to the Resident, Calabar Province, 28 January 1932.
35. NAE UDDIV 2/7/1, File no. EN 78/1931.
36. NAE UDDIV 2/7/1, File No. EN 78/19/31, 'Illicit Distillation of Spirits'.
37. NAE RIVPROF 2/1/17, File no. 68 Vol. III, 'Illicit Distillation', Senior Resident, Owerri Province to District Office, Degema, 27 October 1940.
38. NAE UDDIV 2/7/1, File no. EN 78/1931, 'Illicit Distillation of Spirits.'
39. NAE ABADIST 14/1/169, File no. 275, S.T. Maolu, to District Officer Aba, 8 December 1932.
40. NAE RIVPROF 2/1/15, File no. 68 Vol. 1, 'Illicit Distilling of Liquor and being in possession of Still,' Commissioner of Police, Owerri Province to the Resident, Owerri Province, 25 August 1931.
41. NAE UDDIV 2/7/1, File no. EN 78/1931, 'Illicit Distillation of Spirits.'
42. Ibid.
43. NAC, CADIST 3/1/156, File no 11/31 'Illicit Distillation of Whisky and Spirits,' The Secretary, Southern Provinces, Enugu to the Chief Secretary to the Government, Lagos, 1 October 1931.
44. See, for example, NAE ABADIST 14/1/169, File no. 275, 'Illicit Liquor,' Assistant commissioner of Police to District Officer, Aba, 9 March 1932.
45. NAE RIVPROF 2/1/16 File No. C 68 Vol. II, 'Conviction for Illicit Distillation,' Secretary Southern Provinces to the Resident, Owerri Province, 6 May 1936.
46. See for example, NAE ABADIST 14/1/169, File no. 275, 'Illicit Distillation Returns for the Period January–June 1939,' The Police Headquarters, Aba to District Officer, Aba, 4 July 1939.

47. NAC CADIST 3/1/156, File no 11/31, 'Illicit Distillation-Raid at Ndon Ebom', District Officer, Uyo to Resident, Calabar, 6 May 1932.
48. NAE RIVPROF 2/1/17, 'Illicit Distillation' District Officer, Degema to the Resident, Owerri Province, Port Harcourt, 30 December 1938.
49. NAC CADIST 3/1/156, File no 11/31 'Illicit Distillation', Resident Calabar to Secretary, Southern Province, Enugu, 20 May 1932.
50. NAE ABADIST 14/1/169, File no. 275, 'Minute of a conference with the Residents, Eastern Provinces at Government Lodge, Enugu,' 13 December 1948.
51. The Eastern Regional Committee of the Christian Council comprised representatives of the Church of Scotland, the Methodist Mission, the Qua Iboe Mission, the Church Missionary Society and the Niger Delta Pastorate. See NAE RIVPROF 2/1/17, Eastern Regional Committee to the Resident, Owerri Province, 29 January 1944.
52. Ibid.
53. NAE RIVROF 2/1/17, 'Record of Meeting'.
54. Ibid.
55. NAE RIVOROF 2/1/17, 'Illicit Distillation' Resident, Rivers Province to Secretary, Eastern Provinces, 23 December 1947.
56. Ibid.
57. NAE RIVPROF 2/1/17, 'Illicit Distillation', Resident, Rivers Province to Secretary, Eastern Provinces, Enugu, 23 December 1947.
58. Ibid.

8
Empire and Excise: Drugs and Drink Revenue and the Fate of States in South Asia

Marc Jason Gilbert

Introduction

As the Anglo-Chinese Opium Wars demonstrate, the commerce in mood-altering substances played a major role in the creation of European land empires in Asia. However, this chapter will argue that such trade was also the source of imperial dependence, instability and even decline. This is particularly true of British India. While the opium trade helped to give shape to both the physical and fiscal structure of the Raj, there developed from the late eighteenth century onwards a reliance on excise duties levied on the sale of opium, alcohol and related intoxicants to fund the colonial state. This led, in the later nineteenth century, when political conditions at the imperial metropole facilitated an assault by Radical-Liberal Members of the British Parliament on these 'immoral' sources of revenue, to a division between Her Majesty's Government in London and the Government of India. This division was so serious an internal problem that it led the Government of India to seek independent representation on international bodies governing intoxicating substances: a clear sign of imperial entropy. More significantly, Britain's chief rival for political power on the subcontinent in the twentieth century, the Indian National Congress, was radicalised by the arguments over excise policy aired by its Liberal supporters in Parliament and the self-appointed Indian Civil Service 'guardians' closer to home. This chapter will trace the development of excise policy in the latter half of the nineteenth century in colonial India and will explore the origins and outcomes of the controversies generated by that policy in the 1890s. It will conclude by arguing that the taste for revenues from mood-altering substances that developed during British rule has left post-colonial states in a dangerous condition of dependency on taxes from drugs and drink.

Liquor, opium and the evolution of Indian excise policy

Mood-altering substances have always had a place in South Asian civilization. The use of such substances is discussed in some of the most memorable shlokas of the Rig Veda. Tenth-century Shavite tantric adepts employed alcohol in their caste-bending practices. Yet, despite its association with the behaviour of the lower castes, fermented products such as toddy were widely consumed long before the first modern European distillery was opened on the subcontinent in 1805 by Carew & Company in Kanpur to provide rum to the British army in India. Subsequently, high-quality 'foreign-style' liquor became popular among the well to do, while cheaper and often illicit European-style alcoholic beverages competed with indigenous brews for mass consumption. Hobson-Jobson, the contemporary compendium of Anglo-Indian lore, found mixed evidence that British management of excise licenses and the production of intoxicants (*abkari*) had resulted in an increase in public drunkenness and addiction. However, its authors found it hard to argue that British rule did not increase the means to acquire liquor. They posited an argument which, in today's laissez-faire terms, would read 'the drinking and taxing of liquor is not harmful; the abuse of liquor by drinkers is'.[1] Opium use and production in India and the participation in the opium trade by Indian merchants, notably by Parsi families in Bombay, preceded British rule. Yet, as with liquor, it was hard to argue that any increase in opium consumption thereafter was unrelated to British dominion over India, since opium production and trade contributed so greatly to the rise and expansion of their Raj, as is discussed elsewhere in this book.

By the late nineteenth century the money derived from taxes on the sale of mood-altering substances amounted to approximately 15 per cent of the total revenue of the Government of India (GOI). It was an essential revenue stream. During this period, one of repeated famine, taxes on land were inelastic. Both the land tax and the excise tax on salt were already so high that they were, according to Mike Davis, a cause of the famines themselves.[2] Together, these levies barely covered the expenses of the GOI on its military establishment and the Home Charges, remittances in gold to Britain for expenditures on pensions and purchases there. As the value of the silver-based Indian rupee, like all silver-based currencies, was then dramatically falling, from 2s to less than 1s. 6d, against the value of gold, the GOI was in severe financial difficulties. Despite the GOI's desperate appeals for relief, the British government in London would not tolerate a reduction of the Home Charges or permit India to develop countervailing import duties, as both steps would protect and stimulate competing Indian industry and trade. Unable to raise land and other excise revenue, the GOI was under strong fiscal pressure to expand sales, and hence revenue, of alcohol and opium in times of crisis.

One such crisis arose out of imperial concerns raised by the advance of Russia towards India in Central Asia between 1870 and 1895. The Conservative Party supported an aggressive response to this advance that ultimately led to a war in Afghanistan in 1879. Since that party was particularly opposed to the imposition of an Indian import duty or other external taxes, the Conservative Viceroy, Lord Lytton (1876–1880), sought additional revenues by enhancing excise collections on liquor and opium sales. As a result, revenues derived from liquor skyrocketed, rising from approximately £1,250,000 in 1870 to £2,840,000. The effect on Indian nationalist sentiment, already exacerbated by Lytton's forthrightly racist and heavy-handed administrative policies, was equally dramatic.[3] Many of the earliest Indian nationalists were, like Keshub Chunder Sen, genuine devotees of both Western and indigenous notions of temperance. But genuine prohibitionists or not, they were eager to exploit Lytton's revenue reliance on intoxicants as a means to challenge the moral basis of British rule.

This emergent nationalist sentiment did not escape the Liberal Party, whose victory in 1880 over Lytton's Conservative Party led to campaign pledges to pursue an Indian policy more responsive to India's interests and its peoples' 'reasonable' political aspirations. Lytton's successor, Lord Ripon (Viceroy, 1880–1884), sought to make partners of India's nationalists rather than to estrange them. He began by handing over many of India's municipalities to Indian control and supporting the right of Indians to legal equality as British subjects. These efforts were ultimately eviscerated by the racial divisions between Indians and the British that Lytton's policies had broadened: a 'White Mutiny' arose among Britons resident in India to block such political devolution.[4] However, before this backlash had occurred, Ripon managed to affect a complete overhaul of the methods and practices of liquor and opium policy, which were subsequently endorsed by successive Secretaries of State for India. Ripon sought the maximisation of such revenue from the minimum amount of consumption, principally through stricter enforcement and better administration, which set the pattern for all subsequent excise policy on mood-altering substances. The latter eventually reached such heights of efficiency that collection costs often were only 2 per cent of the revenue gathered. Fees on distribution and taxes on consumption of alcohol and opium were ultimately set, insofar as the GOI deemed practicable, at the highest price that could be obtained short of the threshold beyond which it would encourage widespread illicit manufacture and smuggling. A laissez-faire government like the British Raj, ostensibly operating under libertarian ideals of individual freedom and responsibility, could do little more. Its administrators were confident that any attempt at prohibition would fail due to India's size, the difference between administrative form and style in British territory and Indian States and the differing attitudes about drink and opium among India's diverse cultures.

This change in abkari policy had a slow but significant impact. By 1887 the amount of opium and alcohol consumed by Indians seems to have sharply decreased while revenue generated from both steadily grew. Within a short time the spike in consumption resulting from Lytton's policies had been reduced to pre-Lytton levels. The formal enquiry of an Excise Committee in 1905 found that from 1884 to 1904 liquor sales declined by almost 20 per cent while taxation per gallon almost doubled. The latter led to concerns voiced by a doyen of Indian administration, John Strachey, that liquor prices were so high as to stimulate tax evasion and the use of cheaper illicit drugs, 'especially hemp'.[5]

Steady efficiencies in abkari administration produced so such money at so little cost that, out of liquor revenues alone, the GOI was able to weather the loss of opium revenue brought about by the end of the opium trade and the criminalisation of opium consumption in British India in 1920. The GOI was proud of this achievement and it was noted as early as 1889 that, while it was an alien and authoritarian government, it had met the same standards for the derivation of revenue from the consumption of intoxicants then employed by the world's most temperance-driven democratic republic, the United States of America. There liquor prohibition was introduced only in 1920 and repealed in 1933; though opium importation and its use were controlled between 1909 and 1914, it was not until 1925 that the sale and consumption of opium was criminalised in the United States.[6]

Of course, this 'achievement' won few accolades from the leaders of India's nationalist movement. As Ripon himself feared, Britain had waited too long to effectively co-opt that group. British-sponsored political reforms were destined forever to be 'too little, too late'. His own municipal and abkari reforms served as a case in point. While

Ripon's municipal reforms held the promise of self-government, Indian nationalists berated the administration for not permitting them to exercise control over the licensing of out-stills, the chief source of illicit alcohol production by Indian entrepreneurs, despite the powers granted them under the provisions of law permitting them to exercise 'Local Option' or control of such licenses.[7] It thus mattered little that the evils of the out-still system were being gradually curtailed or that this system had largely been replaced by the much better managed central distillery system in all but rural Bengal, remote areas of the Central Provinces and a few districts elsewhere. Indian newspapers and Congress leaders held that these conditions and the slow pace of abkari reform warranted their accusation that the GOI was deliberately encouraging drinking or at best failing to discourage the habit in order to maximise excise revenue, thus encouraging behaviour which tended to undermine 'the moral and physical fabric of the nation'.[8] The issue, of course, was not so much administrative, as political, and not so much about temperance as about independence. Even though the Indian National Congress was later to admit that the GOI's effort to reform abkari administration was praiseworthy, the abkari policy of the Raj was never beyond cavil. As in the days of Lytton, attacks on the shortcomings of that policy remained an effective means by which India's nationalists could claim that they, not the British, had the greater commitment to the moral as well as material progress of the Indian people.

Between Parliament and the people of India

Indian leaders were initially delighted that Radical-Liberal Members of the British Parliament shared their concerns over abkari affairs. Between 1886 and 1889, Radical-Liberal Members of Parliament Samuel Smith and William S. Caine were feted in India when these temperance campaigners arrived to look into effects of alcohol consumption. Indian leaders also regarded the parallel interest of Caine and his Radical colleagues in opium consumption, Caine and Smith being leaders of the British Society for the Suppression of Opium, as useful to the nationalist cause. However, most Indians soon found that, insofar as abkari policy was concerned, the work of these parliamentarians bore little promise as a vehicle for Indian uplift. They were also to discover that, for many of their Radical Liberal friends, careerism and domestic politics took precedence over the cause of temperance. Indeed, even on the cause of temperance these would-be allies of India had not the least concern for what most nationalists saw as their country's true interests.

The growth of the Radical-Liberal interest in Indian excise issues was rooted in internal Liberal Party politics, which had in the 1880s taken a deeper interest in Indian affairs. Due to Irish obstructionism, Radical backbenchers found it almost impossible to secure time on the floor of the House of Commons with the result that they excluded from the arena in which they could best establish their claims for future advancement and were deprived of the opportunity of bringing before the House those matters with which their constituents were chiefly concerned,[9] such as temperance and the fight against opium consumption. The 'younger' generation of Radicals was worst hit. Excluded from the Cabinet by Gladstone's continued adherence to 'Peel's rule',[10] they had little choice but to resort to every conceivable expedient to gain the attention of the House, including 'snap' votes and midnight divisions which were often so poorly attended that the ruling party rarely took cognisance of the result. In the past it had occurred to these members to seize upon imperial issues whose import would secure openings for the making of motions and the conduct of unlimited debate. However, this was rendered impractical by the divisions within their own ranks between Free Traders and protectionists, Unionists and Irish Home Rulers and adherents of either Charles Dilke's vision of a greater Britain or John Bright's so-called 'Little Englanders'.

India proved to be a different matter. It was significant enough an issue to warrant the attention of a full House, yet sufficiently removed from the concerns of the front bench as to permit a free hand in debate. Radicals had good reason to think that India might offer the opportunities they sought. In 1886 Dadabhai Naoroji, eager to join their ranks in Parliament, had established himself as the agent for Congress in England. This secured an important link between the Radicals and Congress leaders with their wealth of information and catalogue of issues with which to cudgel the Conservative Government on the floor of the House. Moreover, with the improved access to India provided by the recently established regular steamship service through the Suez Canal, Radical members could themselves conveniently sojourn to India for information gathering visits during the long winter recess. In time, Radical Members of Parliament such as Samuel Smith (Flintshire) and William Caine (Barrow-in-Furness), embittered at the short shrift given by Gladstone to the British temperance initiatives they favoured, realised that India offered a large venue for their temperance campaign. Samuel Smith first took a trip to India in 1886, while in the following year William Caine too visited India. Upon his return, he and Samuel Smith threw their weight behind colleague John Slagg's short notice resolution condemning Lord Cross' recent expenditures on strategic railways and merged it with a resolution of their own censuring what they believed to be a conscious attempt on the part of the Government of India to increase the sale of intoxicating liquors for revenue purposes. The combined motions, brought before the House of Commons on 25 March 1888, accused the Government of India of purposefully encouraging the consumption of liquor to pay for the increase in military expenditure caused by such frontier defence operations as the Sind-Pishin Railway extension from Quetta to New Chaman, the linchpin of India's new strategic railway system and a key element of the GOI's frontier policy.

Armed with unofficial data gathered by Caine in India, he and Smith were able to score point after point off John Gorst, the Parliamentary Under-Secretary for India, who could not refute Caine's excise figures without reference to the GOI. So strong were the GOI's arguments regarding the necessity for improved means of communication with the Afghan frontier that Gorst was able to make some telling replies of his own, but so great was the attractiveness of a temperance vote that offended neither British drinkers nor British brewers that Slagg's motion would have passed except for the intervention of W.H. Childers and Randolph Churchill. The former Liberal Secretary of State for War and the former Conservative Secretary of State for India succeeded in convincing the House of the evils inherent in rousing strong pro-temperance sentiment in support of a motion that not only condemned the Government of India without any hearing or inquiry other than a debate on a motion of which inadequate notice was given, but which also censured the GOI for expenditures on military preparations which had been previously sanctioned by both the Liberal and Conservative parties while in power.[11]

However, the Churchill and Childers arguments regarding fairness and imperial security made no impression on Smith and Caine. While their passion for the temperance cause was real it also served other ends. When Caine was given the opportunity to ask for papers to be printed for Parliament on akbari policy, which he knew to be accurate and likely to refute his arguments, he declined to do so. In this, his actions seem similar to those in his dealings on cannabis, where he stands accused of taking an interest simply in order to employ another stick to beat the British government on its imperial policies.[12] The GOI's investigations found that Radical Liberals, including the 'Member for India' Charles Bradlaugh, were using India for personal political purposes. Bradlaugh took a much-needed campaign contribution from interested parties in India, whose case he advocated before Parliament.[13] That

Caine and Bradlaugh may have sincerely believed in the righteousness of the temperance cause in India is not in doubt. That they had supplementary reasons for advocating that cause seems to be equally true.

Slagg's motion served as a wake-up call for the India Office. It was apparent that with snap votes, midnight divisions, the combination of motions, and other Radical tactics the Parliamentary Under-Secretary and his surrogates would need to be prepared for questions designed to make political capital out of the minutiae of Indian administration. A self-proclaimed opponent of the infusion of party politics into the management of Indian affairs and keenly aware of the impediments to rapid official communications with India, Lord Cross was in near panic over the 20 Radical-sponsored and nationalist-inspired short-notice questions and resolutions on India in the years 1886–1888. Cross wrote that it is as if 'you could ring a bell and expect the Government of India to answer and bring up coals at your bidding at a moment's notice. We have democracy with a vengeance.'[14] However, while Cross was convinced that Parliamentary interference in Indian affairs was perhaps a 'fatal thing',[15] he also recognised that the Indian authorities had no choice but to frankly 'accept the situation' and do their best to minimise its dangers by keeping the Home Government fully appraised of policies and events which might attract the interest of Parliament.[16] Cross admitted that 'no one can ever say what the House of Commons will talk about, or what they will leave alone'.[17] In the winter of 1888 John Gorst warned the incoming Viceroy, Lord Lansdowne, that if the Secretary of State was left uninstructed or misinformed, the administration of the Government of India would be 'at the mercy of any ignorant impulse which may seize upon the House of Commons at any moment and your policy is liable to be interrupted and reversed when you least expect it'.[18]

The new viceroy had closely followed the debate over Slagg's motion before taking up his appointment. He was no stranger to Parliamentary manoeuvres or to Indian nationalism, not least of all because his family had a strong history of contact with the Raj; the first Marquis of Lansdowne was regarded as something of an India-expert in Parliament, and the third Marquis was an Indian reformer and also friend of Romohun Roy, the father of Indian nationalism who died at Bowood, the Lansdowne country estate. Before the viceroy-select departed for India he met with Lord Ripon to be briefed on Indian affairs and, most particularly, on the rise of Indian nationalism. This was a cause he came to fully appreciate and whose expectations for political change on the subcontinent he was ready to meet, though he intended to avoid the racist backlash among Anglo-Indians and in Parliament that had eviscerated Ripon's reforms. He knew that the chances of pleasing both the House of Commons and the Queen's Indian subjects were small and observed sardonically that between the two 'the Indian government must have an agreeable time of it'.[19] Nonetheless, he was determined to show the Indian National Congress that, despite its alien character, the GOI was committed to the administration of the country for the benefit of the Indian people and in accordance with their sentiments. He began his administration with a successful effort to repeal the British duty on Indian Silver Plate, turned his predecessor's conservative Indian Councils reform scheme into the beginning of India's modern electoral system in the Indian Councils Reform of 1892, formally recognised the Indian National Congress as a legitimate political organisation and prohibited officials from interfering with it. He also prevented Lancashire from imposing factory legislation on India that was intended to stifle India's industrial growth.[20]

Abkari policy on trial

The abkari issue gave the new viceroy his first opportunity to demonstrate the depth of his commitment to these principles insofar as they could be applied to financial questions. No sooner had he arrived in India than the fourth meeting of the Indian National Congress passed a resolution alleging that a serious increase in the consumption of spirituous liquors had taken place under the existing excise system and urging the Government of India to improve the system so as to discourage insobriety.[21] Lansdowne took note of these suggestions and ensured that his government's current objects were identical to those indicated in the resolution.[22] He also extended every hospitality to William Caine when the temperance advocate returned to India during Parliament's winter recess of 1888–1889 to attend the annual Congress meeting and strengthen his links with the Indian temperance movement. In the first week of January, Caine was the Viceroy's guest at Government House where the two held a series of discussions on the excise issue.[23] At that time Caine expressed the fear that the GOI was encouraging the sale of intoxicating beverages in an attempt to boost revenues without concern for the welfare of its subjects. In support of these arguments he pointed to excise officials who boasted about the increase in revenue they had gained in recent years.[24] Caine's solution to these problems was to have the government create and enforce a uniform all-India *abkari* system that would be managed with a view towards placing morality over revenue and would be based upon local option, a technique which he then acknowledged to be coming into greater use by the Government of India.[25]

The Viceroy agreed with Caine that a reform of the GOI's excise policy was necessary, particularly in Bengal where the out-still system was most prevalent and under attack by the nationalists for lowering the price of liquor.[26] However, he questioned whether human ingenuity could devise a system which could be applied indiscriminately to the whole of India. He also found it difficult to reconcile the local option system with the uniformity of administration that Caine was so anxious to see imposed across India. As far as Caine's criticisms of the existing system were concerned, he assured Caine that he had no objection to the subordination of revenues to morality in abkari policy, since that was a principle which the GOI has always been ready to admit. He demonstrated to Caine that the latter's evidence to the contrary consisted of merely 'the incautious statements of a few local officials who have been anxious to advertise the credit due to themselves for the improvement of the revenue they derived from this source' which was usually due to their efforts at decreasing smuggling and thereby raising the consumption of taxable product.[27] But he also noted that the Bengal authorities were prepared to admit their failings with the out-still programme.[28] He assured Caine that his administration would take immediate steps to redress these deficiencies, extend local option and abolish the out-still system when and whenever possible. On Caine's departure, the Viceroy wrote to John Gorst notifying him that if he was called upon to defend the Government's policy it would be 'a great mistake to represent the existing system as incapable of improvement'.[29] He should 'on the contrary' welcome fair criticism, while at the same time 'dwelling upon the immense difficulty of preventing the abuse of any system however carefully contrived'.[30] The Viceroy also informed Cross that he was re-writing two despatches on the subject, among the largest despatches ever written on domestic policy in India, for use by the India Office so as to provide the Secretary of State with the most up-to-date material for use in Parliament.[31]

At the end of April 1889 Samuel Smith served notice of his intention to resume the Radical assault upon Indian abkari policy. However, the patient and responsive manner in which Lansdowne had treated Caine's suggestions initially appeared to have stolen the Radicals' thunder. On April 5, Cross informed Lansdowne that Caine had written to him that he was satisfied that Lansdowne's government was 'quite alive to the existing evils' and was also determined to apply a remedy. 'I congratulate you', Cross wrote, 'upon having convinced a very obstinate man'.[32] Lansdowne was both pleased and surprised to hear of Caine's disposition. Though they had parted on excellent terms, Caine's last speeches in India appeared 'bitter and unjust' and 'scarcely consistent with his present admission'.[33] In fact, so unjust were Caine's assertions that Lansdowne hastened to provide the India Office with corrections of certain alleged misstatements of Caine's regarding the number of stills and distilleries in Lucknow that the touring Member of Parliament had circulated to Indian newspapers. Caine stated that there were 36 distilleries and 201 liquor shops in the City of Lucknow with an area of 13 square miles and a population of 261,303, but he had conflated these figures with those of the district of Lucknow which had an area of 989 square miles and a population of 696,824. The city had one distillery and 30 stills. The district had no out-stills.[34] Since the Viceroy's staff had gone over such figures with Caine, the error was regarded as unpardonable.

The Viceroy was not alone in his concerns over Caine's public statements. As much as nationalist leaders warmed to Caine's inflammatory anti-government remarks, they grew concerned over their content. Caine seemed to have no idea of the conditions under which most spirituous liquor was consumed: he thought toddy was the source of all evil and responsible for much public drunkenness. In fact, Caine thought that toddy was as potent as fortified ales and was willing to believe it tasted like, and was as intoxicating as, wine. Yet, as the *Jame Jamshed* newspaper proclaimed, 'Toddy is an innocent juice' that should be 'allowed free use'.[35] The *Rast Goftar* declared that 'toddy is the national drink of India. It is admitted to be healthy'[36] reflecting the views of many Indian vernacular newspapers which denounced what they saw as Caine's wrongful imposition of Western attitudes to drink. While some pro-Congress papers continued to voice the Congress view that 'it was most improper for a government to cause the ruin of its subjects for the sake of revenue', they differed with temperance-minded high-caste Hindu Congressmen by declaring that they 'would not go so far as Caine'.[37] Even stalwart pro-temperance newspapers such as the *Sahachar* considered Caine's determination to force a uniform system of alcohol control on India as absurd and, like the Viceroy, considered this approach as hostile to the democratic principle of 'local option'. Worse, from their point of view, Caine's proposal seemed to threaten the non-imported liquor businesses that were largely in the hands of the same Indian middle class that made up or lent support to the Indian National Congress. In such matters the Congress saw itself as serving both its own and the public's best interest by preferring policies that protected the morals of Indians, but not to the point that it deprived them of their customary practices and protected them out of their livelihoods.[38]

The test of Caine's intentions came on April 30, 1889, when Samuel Smith moved a resolution condemning the excise policy of the GOI and alleging that the system afforded great facilities for the spread of drunkenness among the people. Caine then rose in his place and denounced the GOI, making liberal use of the figures the Indian authorities had demonstrated to him were patently in error. He effectively exploited the braggadocio of the revenue officials of the Northwestern Provinces and Oudh who enigmatically spoke about 'the number of shops in three districts being brought up to the Government Standards'[39] even though, as the Viceroy had explained to him, these statements signified that the licensee met the requirements intended to discourage illegal production. Caine's associates in Parliament took the same course. Sensing an opportunity to secure a safe temperance vote and finding the Conservative Whips off guard, the House of Commons threw its support behind the resolution which held that the fiscal system of the Government of India led 'to the establishment of spirit distilleries, liquor and opium shops in large numbers where till recently they never existed [and] increasing consumption' and that the Government of India should take immediate steps to abate these practices.[40]

Cross was incensed at this action but helpless to prevent it. He recognised that the testimony of officials and the phrase 'Government Standard' in no way bore the interpretation placed upon them in debate but 'in unscrupulous hands these passages were made to tell with considerable effect'.[41] Other than cautioning the Lieutenant-Governor of the offending province to carefully watch the 'sayings and doings' of his subordinates, there was little Cross could do to implement the resolution with which, in his opinion, 'the present Government was in complete concurrence'.[42] 'The House of Commons and the Whips owe you an apology,' Cross wrote, 'so does Mr. Caine. The old saying is true—a man convinced against his will is of the same opinion still.'[43] The Secretary of State did his best to soften the blow of defeat. He assured Lansdowne that such snap votes had no real significance and, in his despatch forwarding Smith's resolution to India, he made it clear that the GOI's policy was in no way contradictory to that desired by Parliament.[44] Lansdowne greatly appreciated this gesture and responded with a massive despatch that dealt with the charges brought by the House and laid the foundations of a general reform of the excise system.[45] It indicated that the out-still system had been abolished and the central distillery system extended in Bengal, the principles and spread of local option were now more fully incorporated into general policy, and other measures had been taken to limit sales abuses and the spread of drinking. These efforts were acknowledged and even welcomed by the Indian National Congress, as it much preferred the GOI to direct such reforms than its erstwhile allies in Parliament.[46]

Caine had no intention of allowing the true official GOI's figures to be publicised so soon after the late debate. Knowing this, the India Office staff in London was not surprised when Caine declined the opportunity to have the GOI's recent abkari despatches printed and laid before the House. Though Caine thus insulated himself from criticism, the contents of these despatches eventually became known and silenced criticism in Parliament of the Government's management of spirituous liquor.[47] However, the Viceroy did not let the issue rest without remarking upon the unpardonable carelessness of the Whips that permitted a division to be snatched 'on so important a subject' and the evil results of the Government's defeat;

> A vote of censure carried under such circumstances may no doubt have no significance in the eyes of the English public. It will however be regarded here as a very grave matter, and as a formal condemnation of the Government of India by the English Legislature at a moment when we were making great efforts to remove any imperfections which experience has disclosed ...[48]

Cross sympathised with the GOI, but there was little he could do to prevent Caine from engineering a repetition of the liquor imbroglio over India's opium revenue. At least, on this occasion, the once-bitten GOI and Indian nationalists would begin as allies.

Opium policy on trial

The opium revenue was the third largest source of income for the Government of India during the nineteenth century. During 1891–1892, the combined net receipts from the sale of opium for export to China and from consumption inside the country amounted to £3,844,000 or Rs. 6,150,000.[49] Like excise revenue, however, it was a subject of considerable controversy. Since the 1870s some Indian leaders, including Keshub Chandra Sen and Dadabhai Naoroji, had condemned the China trade as 'a sin on England's head and a curse on India for her share in being an instrument'.[50] They also looked with disfavour upon the use of the drug in India[51] though, as John Richards has argued elsewhere, opium use was relatively benign and considered by most Indians to be a homeopathic treatment for stomach and other aliments used after the fashion of the ancient Greeks and most societies since.[52] Naoroji's anti-opium commitment is noteworthy as he was a Parsi, a community that had been a major force in the opium trade.[53]

As in the case of the abkari campaign, Congressmen like Naoroji and his temperance-minded colleagues had initially worked cordially with critics of opium traffic and use, such as Joseph Pease, William Caine and Samuel Smith. However, as the Society for the Suppression of the Opium Trade's agitation in Parliament got under way in the late 1880s, most Indian leaders and the majority of the Indian-owned press took alarm at the manner in which the Society ignored both Indian interests and the issue of the enormous revenue loss to the Government of India. Directly addressing the 'immorality' of the opium trade that ostensibly was behind this abolitionist effort, Indian newspapers declared that to abandon opium as a source of revenue on moral grounds and then 'make good the gap by ruinous taxes wrung from the lifeblood of an already impoverished people would...be in the highest degree immoral'.[54] Indian leaders saw no reason why, if the British people were genuine in their regard for the Chinese and Indian people, they should not persuade their own government to contribute to the recovery of the losses by India incurred by the abolition of the trade.[55] The Indian-owned Press, which had long cast the British anti-opium crusaders as a 'coterie of extra-superfine moralists'[56] and as a 'powerful class of well-meaning but ignorant fanatics',[57] immediately began to question the intelligence and/or integrity of Caine and Smith, upon discovering that they were not willing to protect Indian interests via compensation.[58]

The waning support of Indians had no effect on the Society's members who sent a delegation to the India Office with a memorial to the British Government which demanded that an immediate effort be made to close licensed opium dens throughout India and restrict the sale of opium to that which was necessary for medical use. They supported this demand with a set of figures for the Northwestern Province and Oudh, which showed that the increase in licences was 36 per cent and brought forward other statistics which showed that revenue receipts from this source had increased from Rs. 79,094,520 in 1881–1882 to Rs. 97,065,130 in 1890.[59] Cross assured the delegates that the Government of India was well aware of the evils attendant upon the unrestrained consumption of opium and forwarded the memorial to the GOI for its information.[60] In due course the Viceroy set in motion the preparation of a weighty despatch which would answer the charges aired in the memorial. To facilitate this task, Lansdowne, undaunted by his failure with William Caine, personally communicated with the leading anti-opium Members of Parliament and attempted to translate their criticism and suggestions into the Government's policy.[61]

As such, by the spring of 1891 the GOI had abolished its much criticised 'minimum vend' system, yet another term merely denoting the meeting of standards that were intended to prevent unlicensed production and sale, in Bombay and had prohibited the consumption of opium and its preparation on the premises of licensed shops. It also placed a ceiling on the amount of opium produced for export and took steps to obtain the cooperation of the Indian States in the suppression of illicit traffic in the drug. Additionally, the sale of opium in Baluchistan was brought under government supervision. Finally, the consumption of opium in Burma was reviewed and a system for the registration of users and the restriction of sales was sanctioned. Only in the North-western Provinces and Oudh did the Government maintain the status quo. Auckland Colvin, whose administration was the victim of Caine's accusations of government support for intoxicant abuse in Lucknow, was furious at the reference in the Society's Memorial that condemned its opium policy. After showing that the local authorities had, in fact, reduced the number of opium shops from 324 to 54 since 1883–1884, thereby exploding the Society's charges, Colvin offered the following blistering remark, redolent of all the hauteur that a senior Indian Civil Servant could muster. There appeared to him to be

> little practical usefulness in engaging in a never-ending discussion [of data] with critics who have shown themselves quite incapable of understanding the uses of accuracy; whose subject is not to prove their contentions but to impress the general public, and to that end to declaim and to denounce; and who cannot therefore be expected to forego their methods of controversy before most cogent reasoning or the most unanswerable statistics.[62]

The GOI's despatch heralding the alteration of its opium policy attempted to show that it was taking every step to check and limit opium consumption short of prohibition which, in their opinion, would merely encourage smuggling and force consumers to resort to more harmful drugs.[63] The changes proposed were regarded as highly satisfactory by moderate members of the anti-opium crusade, such as William Fowler, who was pleased that Lansdowne's administration was 'fully alive to the matter'.[64] However, the majority of their colleagues were determined to carry on their campaign until all opium use and commerce in India was eliminated. On April 10, 1891, they succeeded in passing a snap vote in a half-empty House affirming that the system of opium revenue in India was immoral and urging the Government of India to 'cease to grant licenses for the cultivation of the poppy and sale of opium in British India'.[65] The impact of this resolution was muted by the failure of the House to endorse a rider the moderates attached to the motion, which would have made the Imperial Exchequer responsible for the loss of revenue to India consequent to its passage.[66] As it was very unlikely that the rider would ever gain the consent of Parliament, Cross wrote to Lansdowne that he should 'not take the proceedings of the House of Commons seriously'.[67] Lansdowne replied that it was difficult to believe that sufficient care was taken in cases of this kind by party whips to avert a defeat. He reminded Cross that this was not the first time in which the Government of India had 'by similar neglect, found itself placed in a most embarrassing position'.[68] The ultimate consequences of these incidents would prove dire, for 'if such occurrences are repeated, either the Government of India will become impossible or the authority of the House of Commons must be seriously impaired'.[69]

Cross denied that the whips were at fault and, in fact, had done their best to avoid a defeat. The real cause resided in the politics of agitation. The anti-opium party 'had deluged the British constituencies with literature and exaggerated statements', and these in turn 'deluged their members with the same'. This produced great anxiety among Members of Parliament who wished to avoid a confrontation with their constituents. As a result, many of the Conservative government's backbenchers did not return to the House after dinner, and of those that did return 'a very large number walked out of the House and several voted the wrong way, giving a cheap vote, which cost them nothing, to please the agitators'.[70] Cross observed that the unwillingness of the late Liberal Government to say a word in support of the Government of India was both disgraceful and ominous, showing a want of the sense of responsibility that was in his view characteristic of their backbenchers.

The Viceroy subsequently discovered that there was little difference between Conservative and Liberal statesmanship on the opium question. Between April 1891 and June 1892 the anti-opium party grew in strength under Caine's continued hectoring of Indian policy and this forced the Conservative Government to consider the expedient of appointing a Commission of Enquiry in order to avoid further defeats. It continued to struggle to find an appropriate means of providing the House of Commons with information regarding both the consumption of opium in India and the financial and political ramifications of the abolition of the trade.[71] The Conservatives were thereafter defeated and a former Liberal Secretary of State for India, Lord Kimberley, replaced Cross at the India Office. However, the Liberal government took a similar line. The introduction of an anti-opium motion in the House of Commons in June of 1893 compelled Kimberley to support a Royal Commission on Opium as 'the only means of warding off the attack on Indian revenue'.[72] Lansdowne and Kimberley were both deeply concerned over the prospect of losing the opium revenue in the present financial position of India.[73] Their fears were shared by the Indian-owned newspapers, who again declared that India should only agree to the abolition of the opium revenue if the Imperial Exchequer promised to absorb the loss.[74]

The prospects of avoiding abolition looked bleak when Kimberley, under extreme pressure applied by the Society for the Suppression of Opium,[75] appointed William Caine to the commission. Forwarding a report of the 'very painful impression' created in India by his appointment, Lansdowne explained to his Secretary of State that he did not object to the appointment of a known opponent of the GOI's policy, but he wondered whether 'people at home realise the extent to which the selection of a man of this kind weakens the authority of the Government of India';

> He has abused and libeled us in the most unscrupulous manner: he has again and again been detected in making false statements regarding the administration of the country ... and yet he is selected to come out here as a member of a tribunal which, in the eyes of the native community, is appointed for the purpose of putting the Government of India on its trial. This may not be the way in which the matter is looked at in the House of Commons, but it certainly will be so regarded here.[76]

In the event, Caine had to give up his appointment due to a severe and ultimately fatal malady (angina pectoris) and was replaced by Henry Joseph Wilson, a strongly moralistic anti-opium leader.[77] Wilson subsequently refused to sign the Commission's Report of 1895.[78] The GOI had vented its frustrations at what it regarded as the constant misrepresentation of its excise policy by turning the Commission's proceedings into a showcase for its management of opium production, sale and consumption. These proceedings featured a parade of witnesses supportive of the GOI's policies that included prominent Indians such as the brilliant civil servant and future President of the Congress (in 1909), Romesh Chandra Dutt.

The GOI had no qualms about such manipulation. It had to fight to force the Commission to even take evidence in India and found that the Commission's members were so eager to leave India for China that they took little interest in their work. It thus saw no reason not to force what it regarded as the truth down the Commissioners' irresponsible throats. Such orchestration was made easier for the GOI due to the presence on the Commission of Sir James Lyall, a recently retired former Lieutenant-Governor of the Punjab. Lyall had often disagreed with the Viceroy and other Indian officials on questions of state, but he still retained their respect. He, in turn, was mindful of the difficulties that his former colleagues had in making their case before their critics. Much the same could be said for one of the two Indian members of the commission, Laskhmishwar Singh, the Maharaja of Darbhanga, the only major Indian prince to openly support the Indian National Congress who had also proved his loyalty to the GOI through service on the Viceroy's Legislative council. In the end, after enduring a pro-government barrage facilitated by these members, the Opium Commission suggested stricter regulation of opium sales but did not favour the prohibition of the growth or use of opium on the subcontinent. It argued that the Indian people were not prepared to bear the cost of prohibition, and India's finances were not in a position to bear the loss of revenue if the export of the drug, let alone its local consumption, was halted.[79] This pattern, of the GOI successfully defending its liquor and opium policies, extended to related excise policies governing the production of cannabis. This drew questions from Parliament, but resulted in a similarly benign commission of enquiry which validated the GOI's then current practices.[80]

The GOI may have won the battle with what Lord Cross called the 'excesses of democracy' in the 1890s but, as he predicted, the struggle fatally undermined both the GOI's authority and the empire itself. During the Opium Commission proceedings, Indians like Romesh Chandra Dutt assisted the GOI's struggle with Parliament as it coincided with their own interests. But the unseemly conflict between the GOI and Parliament took its toll. Nationalists were angered at Parliament's disinterest in justice for India and were also distressed at the GOI's strident defense of its excise policy which strained the limits of credulity: while local excise officials unwisely boasted of higher abkari revenues to secure promotion there remained the suspicion that they encouraged greater liquor sales to serve that end. Dutt and Dadabhai Naoroji had believed that if the British could only be made to understand their failings as rulers, then the GOI would address them. But the scrap between the House of Commons and the GOI suggested to these observers that Parliament was little more than a venue for harassing their colonial masters and that the latter were interested only in their own narrow agenda.

The outcome of this was that in 1904 Dadabhai preceded his third term as President of the Indian National Congress by demanding 'Swaraj' or self government for India. For his part Dutt, after his retirement from the Indian Civil Service, became President of the Indian National Congress (1899) and an outspoken critic of GOI policy. Other events, such the GOI's response to the return of famine of 1897, may have influenced this change in attitude but it requires no stretch of the imagination to conclude that the excise controversies of 1889–1895 helped engender the more radical views on India's political future expressed by Dutt and Naoroji. The growing conviction of these Nationalists that the Raj was unable to heal itself influenced those who succeeded them, among whom were Gopal Krishna Ghokale and, through him, Mohandas Karamchand Gandhi.

Conclusions

> If I was appointed dictator for one hour for all India, the first thing I would do would be to close without compensation all liquor shops, destroy all toddy palms such as I know in Gujarat, compel factory owners to produce human conditions for their workmen and open refreshment and recreation rooms where these workmen could get innocent drinks and equally innocent amusements... For the loss of revenue from drinks, I would straight away cut down military expenditure.[81]

While Gandhi famously took the above stance on alcohol, the Indian nationalist movement as a whole did not share his dream of uncompensated prohibition. Most were content to limit the liquor issue to swarajist action against foreign imported liquor. This protected Indian entrepreneurs, many of whom were supporters of Congress, who were duplicating the foreign product chiefly for the consumption of the Indian middle class (i.e. to say, the legal production of such liquor). They thus served themselves as they served their country.

The prevailing political pragmatism and moral ambivalence among Indian nationalists towards the control of liquor is reflected in Article 47 of the Indian Constitution. It employs an equivocal paragraph which proclaims the prohibition of liquor consumption as an 'ultimate goal' of state policy but does not prevent the continuation of the entire abkari system of the British Raj.[82] That system has grown from providing about 15 per cent of the revenues of the Raj to providing 24 per cent of the Republic of India's current annual revenue. This may help explain the fate of successive national advisory and planning boards that have tried to limit access to these products. When assisting in preparing India's second five-year plan, the Shriman Narayan Committee held that 'excise duties on narcotics and intoxicants collected in the states are highly iniquitous, regressive and anti-social' and recommended nationwide prohibition by 1958. The Tek Chand Committee of 1964 hoped this goal would be achieved in 1970.[83] Where prohibition has been tried in individual states, it has failed along the lines predicted by the British; it served only to encourage smuggling and to cripple the government's ability to provide adequate public services. But perhaps an even larger problem is that where some control over liquor production has been achieved, workers complain about the loss of jobs.[84] Recently, when women political activists succeeded in closing a brewery in the name of ending public drunkenness and domestic violence, it provoked bitterness among the brewery's workers, one employee complaining that 'the women who are having problems should have control over their husbands... They shouldn't ask to shut down a whole factory that affects all the employees and all their families'.[85]

Moreover, the democratic Republic of India faces the problems of a democratic state when dealing with alcohol. The 'abkari lobby' or 'liquor mafia'[86] exercises control over many Indian political parties via political contributions. The only major political parties not so co-opted include the Naxalites in Andhra Pradesh who took arms against the liquor barons in that state, ultimately forcing them to accept police bodyguards. This led to an alliance between the liquor barons and the state which resulted in the unusual step of liquor being put on sale at police posts.[87] Such alliances are proliferating.[88] The current trend in India towards economic liberalisation may make matters worse as joint ventures between leading Indian distilleries and international liquor giants such as Fosters Group Ltd have added leverage to the Indian liquor baron's effort to crush local rivals and smooth the political resistance that might otherwise attend their monopolistic practices.[89] The globalisation of liquor markets has led to the same displays of ignorance of conditions on the subcontinent that plagued the question over a hundred years ago. Despite his fame as an opponent of alcohol consumption, Gandhi has been included as part of a 'Keep Walking' promotional campaign promoting Johnny Walker whiskey.

While the magnitude of the power now wielded by the liquor barons is a recent development, India continues to suffer the same debilitating debate over the role of excise administration in promoting consumption that helped undo the British Raj. Even the terms of that debate are familiar. Whereas the unabashedly imperialist John Strachey once defended the British GOI liquor taxation system by arguing that alcohol prices should not be set so high as to shift consumption to illicit drugs, some equally self-serving liquor contractors are now urging the Raj's successors to 'reduce the price of liquor to check addiction among the youth [as] cheap liquor will prevent the youth from consuming drugs'.[90] In January of 2002 the government of the state of Maharastra found itself defending an announcement relating to promotions for excise officials and police officers in whose districts liquor sales rose:

> Authorities say the scheme will only increase the efficiency of the officers. State Excise minister Anil Deshmukh said: 'It is funny, but true. Increase in liquor sale is being treated as a barometer to check the performance of our staff... If a particular area registers a hike in revenue from country liquor it means excise officers have controlled bootlegging as more people turn to licensed country liquor shops when they don't get cheap illicit liquor'.[91]

Because opium use was considered relatively benign on the subcontinent, the Republic of India did not rush to re-authorise the 1920 British Indian legislation criminalising opium consumption in India. It also ensured that opium production was protected with the result that post-independent India is the world's leading producer of opium and opium-derived drugs. The Government claims that there is little leakage of this product into India society at large, but evidence is lacking. After 16 tons of illicit opium was seized, the Indian authorities claimed that 6 of the 16 tons of opium seized was from Afghanistan, which does not quite serve to deflect concern as to where the other 10 tons was produced.[92] An opium mafia has yet to emerge in India, and consumption remains relatively low, but international observers have listed opium production and traffic as one of the greatest threats to India's social structure, security and status as a regional superpower.[93]

In Pakistan drugs are an immediate threat to the survival of the state. Just a few years ago Pakistan was a regional power that was master of the Taliban in Afghanistan and equal to any Indian threats over their common claims in Kashmir. It is now in disarray, a condition worsened by opium. In 1999, Pakistan had almost halved the number of acres cultivating poppy but its North West Frontier province is now reverting to poppy production. Moreover, Afghan-produced opium has flooded the country, its own addicts consume 92 metric tons of heroin a year, and drug-related money-laundering is rampant. While Pakistan faces many threats to its security,

> the greatest threat will be the lure of the drug money that may induce the government to support the drug cultivation and trafficking, if not directly, indirectly in many ways. This will further have its impact on the entire society and its various institutions. For example, the government may be forced to depend dangerously on the drug money for its economic survival; the drug barons may start wielding enormous power either directly or indirectly and the polity may find it extremely difficult and/or impossible to get rid of the 'drug power'. If that happens, it may be the beginning of an end.[94]

Since liquor and opium use, licit and illicit, have always had a place in the Indian economy, apologists for the British Raj could argue that it did well to control intoxicants in India and that any current problems are the responsibilities of its successor states. Yet the British Raj acted as a vehicle for modernisation in India, a process that included vastly improving the efficiency of the production and distribution of mood-altering substances. Moreover, British Indian administration institutionalised the abkari revenue stream to such an extent that it would have been exceedingly difficult for its successor states to do without it, especially given the challenges of decolonisation, nation-building and global economic liberalisation. The British Empire's taste for excise revenues drawn from the trade in mood-altering substances exposed its institutions as self-serving and blinkered and undermined the moral basis of its rule. It was also the basis for a growing dependency on such revenues that threatens the post-colonial states of South Asia.

Notes

1. Colonel H. Yule and A. Burnell, *Hobson-Jobson: A Glossary of Colloquial Anglo-Indian Words and Phrases, and of Kindred Terms, Etymological, Historical, Geographical and Discursive* (London: John Murray 1886), p. 2.
2. See M. Davis, *Late Victorian Holocausts: El Nino Famines and the Making of the Third World* (New York: Verso 2001).
3. J. Strachey, *India: Its Administration and Progress*, third edition (London: MacMillan 1911), p. 193.

4. See E. Hirschmann, *'White Mutiny': The Ilbert Bill Crisis and the Genesis of the Indian National Congress* (Delhi: South Asia 1980).
5. Strachey, *India*, p. 186.
6. C.E. Barnard, Minute Paper dated June 3, 1890, Revenue and Statistic Department File no. 232/1890.
7. *The Bengalee*, November 19, 1887; *Sanjivani*, April 30, 1887, Bengal, Indian Newspaper Reports, Indian and Oriental Collections, the British Library, London (hereafter referred to as INR). Under the central distillery system, a fixed duty was placed on each gallon of spirit manufactured and issued for sale. Under the out-still system, the duty was imposed not on the quantity produced but in the gross by lump sum payments on the basis of auctions. The central distillery system was clearly the more effective means of checking consumption, but it required a great degree of monitoring which was impractical in remote areas, hence the use of out-stills. See Revenue Despatch From India, no. 29, February 4, 1890.
8. *Sanjivani*, April 7 and 14, 1888, Bengal INR, *Mahatma*, May 17, September 29, 1885; *The Report of the Indian National Congress* (1888), pp. 60–3.
9. D.A. Hamer, *Liberal Politics in the Age of Gladstone and Rosebery* (London: Oxford University Press 1972), pp. 89, 162 and 227.
10. Refers to Robert Peel's habit of reckoning advancement to the Cabinet via seniority.
11. Hansard, 3rd Series, 323 (1888): cols 1093–179.
12. J. Mills, *Cannabis Britannica: Empire, Trade and Prohibition* (Oxford: Oxford University Press, 2003), pp. 96–102.
13. See 'Extract from the Diary of the Resident of Kashmir for the week ending 5th August 1890' enclosed in Lansdowne to Cross, September 1, 1890, British Library (London), Oriental and India Collections, Indian Office Library and Records, European Manuscripts (hereafter referred to as IOL MSS Eur) E. 243/29. Bradlaugh, who confessed to be 'close driven' and in need of funds in July, asked the Maharaja's agent in London, William Digby, for £100 immediately after placing a question on Kashmir before the Parliamentary Under-Secretary of India, Sir John Gorst. After meeting with Digby about the money, Bradlaugh was sufficiently worried over the ethics of this action to insist that Digby not make public any mention of this 'advance' on the campaign support promised by Digby's Indian Political Agency. He told Digby that the 'help' could be open and public, but the 'advance' could not. See Motilal Ghose to William Digby, 27 (May?), 1892, IOL MSS Eur. 769/9; Charles Bradlaugh to William Digby, June 30, July 25, and July (no day provided) 1890, IOL MSS Eur. 769/7.
14. Cross to Dufferin, June 29, 1888, IOL MSS Eur. E. 243/18.
15. Cross to Dufferin, September 11, 1888, IOL MSS Eur. E. 243/18.
16. Cross to Dufferin, June 29, 1888, IOL MSS Eur. E. 243/18.
17. Cross to Dufferin, March 8, 1888, IOL MSS Eur. E. 243/18.
18. Gorst to Lansdowne, December 24, 1888, IOL MSS Eur. D. 588/11.
19. Lansdowne to Godley, March 29, 1888, IOL, MSS Eur. D. 558/11.
20. See M.J. Gilbert, 'Lord Lansdowne and the Indian Factory Act of 1891: A Study in Economic Nationalism and Proconsular Power' in *The Journal of Developing Areas*, 16, 1982, pp. 357–82.
21. Ibid., p. 229

22. *Report of the Indian National Congress*, (1888) Resolution VII, p. 60.
23. Lansdowne to Gorst, January 15, 1889, IOL MSS Eur. D. 558/16.
24. D. Mackenzie Wallace to Godley, February 10, 1889, IOL MSS Eur. D. 558/11.
25. Lansdowne to Cross, January 15, 1889, IOL MSS Eur. D. 588/2.
26. See *Gramvasi*, December 30, 1888; *Sanjivani*, January 5, 1889, Bengal INR.
27. Lansdowne to Gorst, January 15, 1888, IOL MSS Eur. D. 558/16.
28. Ibid.
29. Cross to Lansdowne, April 5, 1889, IOL MSS Eur. D. 558/2.
30. Lansdowne to Gorst, January 15, 1889, IOL MSS Eur. D. 558/11.
31. Ibid.
32. Cross to Lansdowne, April 5, 1889, IOL MSS Eur. D. 558/2.
33. Lansdowne to Cross, April 29, 1889, IOL MSS Eur. D. 558/2.
34. D. Mackenzie Wallace to Godley, February 19, 1889, IOL MSS Eur. D. 558/11.
35. *Jame Jamshed*, July 16, 1889, Bombay Indian Newspaper Reports.
36. *Rast Goftar*, July 21, 1889, Bombay INR.
37. See *Samaya*, January 11 and 25, 1889, Bengal INR.
38. Lansdowne to Gorst, January 15, 1889, IOL MSS Eur. D. 558/16. See also the *Samaya*, January 25, 1888; *Sahachar*, January 16, 1888, Bengal INR.
39. Hansard, 3rd Series 335 (1889): cols 830–7.
40. Hansard, 3rd Series 335 (1889): col. 169.
41. Cross to Lansdowne, May 2, 1889, IOL MSS Eur. D. 558/2.
42. Ibid.
43. Ibid.
44. Revenue Despatch. No. 30, April 15, 1889.
45. Revenue Despatch. No. 29, February 4, 1898.
46. *Report of the Indian National Congress* (1890), Resolution IV.
47. But not of the excise policy. At the insistence of Radical-Liberal Members of Parliament William Caine and Mark Stewart, an inquiry was made into the cultivation of hemp drugs. See Revenue Despatch to India, no. 36, March 16, 1893.
48. Lansdowne to Cross, May 6, 1889, IOL MSS Eur. D.558/2.
49. Kimberley to Gladstone, March 1, 1893, British Library Additional Manuscripts (hereafter referred to as BL Add. MSS) 44229, fo. 79.
50. Dadabhai Naoroji to Louis Mallet, November 16, 1880, quoted in D. Naoroji, *Poverty and Un-British Rule*, p. 215.
51. *Mahratta*, July 17, 1881; *Kesari*, May 2, 1882, Bombay INR; *Sandharani*, May 11, 1889, Bombay INR.
52. J. Richards, 'Opium and the British Indian Empire: The Royal Commission of 1895' in *Modern Asian Studies*, 36, 2, 2002.
53. As in the case of Sir Jamstejee Jeejeebhoy, who garnered a British baronetcy in 1843 after he 'flouted the religious taboo on tobacco and alcohol to amass a fortune from the Opium trade with China. He diversified his interests soon after, and was a shareholder in Bombay's first English newspaper, "Bombay Courier". During the cotton boom he held 13 directorships in various cotton mills'. See his online biography at http://theory.tifr.res.in/bombay/persons/jamsetjee-jeejeebhoy.html. Accessed, 4 September 2003.
54. *Amrita Bazar Patrika*, July 9, 1880.
55. *The Hindu*, December 6, 1890; *The Bengalee*, April 18, 1891.
56. *Gujurat Darpan*, March 18, 1882, Bombay INR.

57. *Mahratta*, March 22, 1885.
58. *Sahachar*, April 15, 1891; *Som Prakash*, June 8, 1891, Bengal INR.
59. Revenue Despatch to India, no. 68, August 25, 1890.
60. Ibid.
61. Note, July 14, 1891. IOL MSS Eur. D. 558/58, William Fowler, March 25, 1891, IOL MSS Eur. D. 558/13.
62. Revenue Despatch no. 94, October 14, 1891.
63. Ibid.
64. W. Fowler to Lansdowne, March 25, 1891, IOL MSS Eur. D. 558/13.
65. Hansard, 3rd Series, 352 (1891): col. 304.
66. Hansard, 3rd Series, 352 (1891): cols 342, 382–5.
67. Cross to Lansdowne, April 17, 1891, IOL MSS Eur. D. 558/3.
68. Lansdowne to Cross, April 15, 1891, IOL MSS Eur. D. 558/3.
69. Ibid.
70. Cross to Lansdowne, May 8, 1891, IOL MSS Eur. D. 558/3.
71. Cross to Lansdowne, July 7 and 10, 1891, IOL MSS Eur. D. 558/3.
72. Kimberley to Gladstone, March 1, 1893, BL Add. MSS 44229, fos 77–79
73. Ibid.; Lansdowne to Barbour, June 20, 1893, IOL MSS Eur. D. 558/6.
74. *The Hindu*, July 3, 1893.
75. Kimberley to Lansdowne, August 18, 1893, IOL MSS Eur. D. 558/6.
76. Lansdowne to Kimberley, August 15, 1893, IOL MSS Eur. D. 558/6.
77. W.S. Fowler, *A Study in Radicalism and Dissent: The Life and Times of Henry Joseph Wilson, 1833–1914* (London: The Epwroth Press 1961), p. 998.
78. For Wilson's able Minute of Dissent see Ibid., pp. 146–168.
79. See Great Britain, Parliament, Parliamentary Papers (Commons) 1890, vol. 81, C.6042.
80. See Mills, *Cannabis Britannica*, pp. 99–123.
81. *Young India*, 1931.
82. According to Article 47, it is the 'Duty of the state to raise the level of nutrition and the standard of living and to improve public health – the state shall regard the raising of the level of nutrition and the standard of living of its people and the improvement of public health as among its primary duties and in particular, the state shall endeavour to bring about prohibition of the consumption except for medical purposes of intoxicating drinks and drugs which are injurious to health.'
83. See 'Prohibition and the State in India,' prepared by the anti-arrack movement at http://www.uohyd.ernet.in/sss/dhistory/arrack/proj_s.html. Accessed, 4 September 2003.
84. Amy Louise Kazman, 'Prohibitionists Gaining in India' at http://www.dui.com/alcoholissues/Alcohol/india.prohibition.html. Accessed, 4 September 2003.
85. Ibid.
86. See Shankar Pratap Singh, 'Transnational Organized Crime: An Indian Perspective', at http://www.unafei.or.jp/pdf/no59/ch29.pdf. Accessed, 4 September 2003.
87. Also mentioned in the above cited 'Prohibition and the State in India', at http://www.uohyd.ernet.in/sss/dhistory/arrack/proj_s.html. Accessed, 4 September 2003.

88. 'Police Helping Liquor Mafia at Vellarada', *The Hindu*, October 1, 2002 at http://www.hinduonnet.com/2002/10/01/stories/2002100108130300.htm. Accessed, 4 September 2003.
89. K.S. Chawla, 'Lobbying on for Liquor Vends', Ludhiana Tribune Online Edition, February 7, 2003, at http://www.tribuneindia.com/2003/20030207/ldh1.htm. Accessed, 4 September 2003. For a roseate view of 'the global consolidation' of liquor distribution, see Boby Kurian, 'High Spirits', Business Line – Internet Edition (produced by the *The Hindu*), December 12, 2002, at http://www.thehindubusinessline.com/bline/catalyst/2002/12/12/stories/2002121200070300.htm. Accessed, 4 September 2003.
90. K.S. Chawla, 'Lobby on for Liquor Vends', *Ludhiana Tribune*, February 7, 2003, at http://www.tribuneindia.com/2003/20030207/ldh1.htm. Accessed, 4 September 2003.
91. 'Police in India to get Cash Bonuses if Liquor Sales Rise', *Ananova* (Online News Service), September 9, 2002, at http://www.ananova.com/news/story/sm_667137.html?menu=Accessed, 4 September 2003.
92. See 'Statement by Ambassador T.P. Sreenivasan at the 44th Session of the Commission on Narcotic Drugs' in Vienna on March 22, 2001 at http://www.indianembassy.at/US/About/Sreenivasan/Speech_2001-03-22_UNDCP.html. Accessed, 4 September 2003.
93. 'India is the world's largest producer of licit opium for the pharmaceutical trade, but an undetermined quantity of opium is diverted to illicit international drug markets. India is a major transit country for illicit narcotics produced in neighboring countries, and a producer of hashish and methaqualone.' See 'India Files', MSNBC News at http://www.msnbc.com/modules/secretempire/india.asp. Accessed, 4 September 2003.
94. D. Suba Chandran, 'Drug Trafficking and the Security of the State: Case Study of Pakistan' at http://www.idsa-india.org/an-sep8-7.html. Accessed, 4 September 2003.

9

Powders, Potions and Tablets: The 'Quinine Fraud' in British India, 1890–1939

Patricia Barton

> The man who withheld the quinine from that tablet or mixture or who allowed it to be labelled 'quinine' when he knew that it was not quinine was as surely the murderer of that unfortunate patient as he would be had he thrust a knife into his back; the accessories are the doctors who did not see that the quinine came from a 'reliable source' and the authorities who make no effort to ensure that every source available to the public *is* a 'reliable source'.[1]

Introduction

In British India malaria was the greatest cause of morbidity, although its impact on mortality levels was disputed. While 'fevers' regularly accounted for the highest proportion of registered deaths, questions were raised regularly about the abilities of the village registrars to accurately record the causes of death. The Sanitary Commissioner of Assam, Carr-Calthrop, refused to outline public health plans based upon local statistics and declared that 'arguments founded on such bases are useless, a mere waste of paper'.[2] This was supported by Rogers of the Calcutta Medical School who was scathing about the official mortality statistics: using his 'gleanings' from Calcutta post mortem reports, he suggested that malaria accounted for only some 5 per cent of deaths.[3] However, he failed to give due recognition to regional variations. While Bengal as a whole suffered badly from the disease, major epidemics of malaria in Calcutta were rare. Indeed, it was on account of his transfer to Calcutta that Ronald Ross was forced to focus more on avian rather than human malaria, while pursuing his research into the mosquito vector, since there were so few local victims, a situation which he regarded as yet one further example of the idiocy of the British administration in India.[4]

Some officials, though, remained convinced of malaria's contribution to Indian mortality. One year prior to Rogers' work, Acton and Knowles argued, 'That malaria is the most important cause of mortality in India is scarcely to be doubted and it is much to be deplored that we know so little about this important aspect of malaria.'[5] In general it was accepted that malaria accounted for some one million deaths per annum, rising to two million in an epidemic year.[6] It was also suggested that death was not the only measure of the impact of the disease. In 1935 a sub-committee of the Colonial Advisory Council of Agriculture and Animal Health estimated that 30 per cent of the Indian population suffered from malaria at any given time.[7] At the same time, Sinton argued that 'the financial losses to the individual and the family alone can be calculated at not less than ... about 80 million pounds sterling per annum'.[8] It was the knock-on effect of this incapacity on the economy that most alarmed the Government of India. Gill had reported earlier in the twentieth century that during the malaria epidemic in the Punjab of 1908, in which 90 per cent of the population of the malaria districts had been infected, 'Labour was seriously disorganised and the price of foodstuffs throughout the province ruled high'. Not only had this increased the cases of secondary infection due to poor nutrition but, with so many infected, services on the North-Western Railway had had to be curtailed causing 'great financial loss not only to the department, but also of commercial and agricultural interests generally'.[9]

It was acknowledged that it would be a major challenge to treat malaria in India, and one report on the Punjab epidemic of 1908 pointed out that '[we must not] forget that in attempting to reduce malaria mortality in India we are proposing to undertake a task infinitely greater than anything of a like sort previously attempted in any part of the world'.[10] However, it was a challenge overshadowed in the colonial priority list by the more dramatic epidemics such as plague and cholera. Arnold has been able to argue that for the period between the mid-1890s and the 1920s 'the significance of the plague for the political epidemiology of colonial India was far greater than that of the concurrent epidemics of malaria or influenza'.[11] In the late 1990s Harrison could claim that this order of priority had been replicated in the post-colonial context.[12]

Indeed, even those measures devised by the colonial state for addressing malaria ended in failure. This chapter will consider one of the key reasons for this dismal outcome. It will argue that a number of policies and strategies on the part of the colonial state, including excise raising powers on intoxicants, provisions for medical training and the intricacies of anti-malaria campaigns, combined to create a context where a black market developed for treatments and prophylactics. This black market resulted in widespread adulteration of cinchona products to the extent that many of those available, either through representatives of the colonial state or at the local bazaar, were either ineffective or positively harmful. The story is an investigation into the knock-on effects of imperial policy on narcotics in colonial South Asia. Despite the availability of a cadre of Chemical Examiners and related scientists in India in the late nineteenth and throughout the twentieth century, they were rarely concerned to check the quality of medicines. The reason for this was simple; their efforts were focused on the more lucrative activity of guaranteeing that high-grade narcotics reached the markets controlled by the British in India for opium and cannabis.

'The quinine fraud'

Quinine therapy has been regarded as 'one of the classical examples of empirical treatment'.[13] Cinchona was utilised as a treatment for malaria centuries before either the aetiology of malaria or the action through which the drug worked was understood. Indeed for much of its history, quinine was utilised as a broad-spectrum anti-febrile rather than as a specific anti-malarial therapy.[14] Quinine therapy had been popularised in India by surgeons accompanying the additional British battalions sent to India when the 1857 Rebellion coincided with an epidemic of malaria. By then the four major alkaloids had been isolated from cinchona bark, quinine, cinchonine, quinidine and cinchonidine, inaugurating the era of scientific medicine. Faced with an increasing problem of fever deaths in the nineteenth century, the Government of India had to act. From 1861 the policy of the Government of India was to make the subcontinent self-sufficient in cinchona products. Plantations were established under the control of the central authorities and the provincial governments of Bengal and Madras.[15] In 1866 a Cinchona Commission was appointed in Madras to decide which of the alkaloids should be adopted as the official standard treatment, a sample of each being sent to different doctors to determine its clinical efficiency. Although quinidine showed better results than pure quinine, it had more side effects, so the Government of India continued to promote the use of quinine sulphate. The breakthrough came in 1888 when government chemists succeeded in extracting quinine from the bark. Factories were established to produce small single-dose packets which could be sold to the general population, the famous 'pice packets', so-called from their price.[16]

Over subsequent decades the self-sufficiency goal was never realised. In the late 1930s, and on the basis of its estimate that annually one-third of India's population was subject to malaria, the Colonial Advisory Council calculated a yearly quinine demand of 1.25 million lbs, effectively the entire annual world production. Russell, the Public Health Commissioner, revised this downwards to 600,000 lbs, while his successor, Cottar, reduced the figure further to 350,000 lbs.[17] Yet the Cinchona Officer for Bengal had noted earlier in the decade that 'the total consumption of quinine in India [has] kept remarkably steady at about 211,000 lbs per annum'.[18] Despite the grand claims made for 'rural mass treatment' schemes, with no province providing more than an average of 10 grains of quinine per person per annum, Russell rightly could argue that the quininisation programmes were 'ludicrously inadequate'.[19]

Yet in the same decade another report made it evident that it was not just inadequate amounts of cinchona products that were reaching the consumer. In 1930 R.N. Chopra was put in charge of a committee that investigated the quality of the cinchona medicines available both through the colonial state and through private channels. This was the culmination of decades of growing concern about the issue of the adulteration of these medicines. During the late nineteenth century it had been taken as an article of faith that any quinine provided in hospitals, dispensaries or official agencies would be 'always as pure as it can be made'.[20] Officials argued that it was only unofficial supplies found in the bazaar which might be adulterated with flour, chalk, weaker alkaloids and other substances. It was accepted that official supplies had to be pure to counter popular suspicions about the quininisation efforts. King viewed the pice packets as 'the crowning advantage of the Government Quinine Factory', part of an ambitious project 'to put the only efficient medicine for the most fatal disease in the Indian Empire within reach of its poorest inhabitants'.[21] If the treatment did not work, 'the cause of the failure is not far to seek – the drug has not been taken at all, the dose has been insufficient, or it has been taken irregularly'.[22] Human error or prejudice was to be blamed and not the drug itself, 'the one efficient weapon to hand'.[23]

Yet, some questions about the official supplies had already been raised. In 1907 Megaw had become suspicious about the number of his malaria patients failing to be cured by quinine. Discounting contemporary interpretation of such failures as 'quinine intolerance' or 'idiosyncrasy', he collected samples of quinine used in the Calcutta Medical School Hospital for analysis. He was astounded to discover that up to 25 per cent of his supplies were under-strength. He reported that the failure of his patients to be cured was not the fault of the quinine but that they 'had failed to receive the doses of quinine prescribed for them'. He concluded, 'In the case of institutions where large quantities of such a saleable drug as quinine are dealt with it is necessary to keep a close check on the stock mixtures and where the drug is given prophylactically the results will often be found to improve marvellously if the acid solution is made up and administered in one's presence.'[24] He recommended the periodic inspections of hospitals and dispensaries. Despite the shocking nature of his report no central action was taken. However, local authorities began to substitute tablets for the powders. The acknowledged benefits were twofold: 'this arrangement has found to be exceedingly satisfactory, as not only is the drug much appreciated in this form by the people, but it ensures accuracy of the dosage and prevents adulteration'.[25] Despite this, more anecdotal reports of suspected adulteration began to appear. In Burma, the Public Health Commissioner reported that the population had accepted quinine as the best anti-malarial, but was put off by the taste. Unfortunately the local authorities did not yet have supplies of tablets to meet demand, and instead patients looked to the local market for their pills. Their purchases, however, contained little quinine and were 'apparently as insoluble as wooden nutmegs'. The MO at Mojok had treated one unfortunate purchaser: 'a patient was relieved of 35 of these tablets only after an enema'.[26]

While such evidence began to filter through, there was still no official action to preserve the quality of official supplies. Although some Provincial Chemical Laboratories had been established and were now testing for the adulteration of food products such as ghee, oils and milk, few analysed the purity of medicinal drugs. Yet popular belief in the purity of official quinine supplies was also wavering. The Chair of the First Malaria Conference in 1909, Sir Herbert Risley, spoke of rumours about the pice packets: 'people said absurd things, one of which is that postmasters would take the trouble of opening several thousand packets – not easy packets to open by the way – and that they would extract the quinine, would sell it and would substitute something else'. He was quick to deny any such charge.[27] Shortly afterwards, at the Second All-India Sanitary Conference in Madras, Dr Nair was to have presented a paper entitled 'Inspection of Food and Drugs in relation to Public Health' at the Hygiene Section. However, as he explained, 'On closer inspection I found that nothing had been done in the way of inspection of drugs, not only in Madras but in the other big cities in India also.' He argued that India required a Pharmacy Act 'to meet the difficulties, which we at present are confronted with in enforcing the purity of drugs in the interest of public health'.[28] While Nair's views received general support in the wider discussion amongst delegates, no resolution was made on the issue at the end of the conference.

In 1928 Megaw returned to the subject, overseeing a wider analysis of 100 hospitals and dispensaries in Bengal and Assam in which the rate of under-strength tablets was again roughly 25 per cent, with 8 per cent being over-strength. He believed that 'The dispensaries from which the stock solutions were taken are not likely to be worse than the average.' All had been under the charge of a Medical Officer, though Megaw's team noted that some had been less than enthusiastic about taking part in the tests. While they visited dispensaries without prior warning, they accepted that word would quickly spread throughout the district when they arrived. This made the high level of adulterated preparations all the more worrying. Megaw recommended governmental spot checks to control the level of adulteration and a simple chemical test which Medical Officers could conduct to test the strength of the quinine preparations. The report concluded:

> Many observations on the clinical action of quinine are utterly vitiated because the investigators have assumed that writing a prescription or giving an order constituted evidence of the administration of the stated amounts of quinine. The truth is that we have no assurance that quinine has been taken unless we have actually seen the mixture made up and administered.[29]

This finally shocked the Government of India into action. As well as in state hospitals and dispensaries, Megaw's tests had been conducted in those run by tea plantation and mining companies in Bengal and Assam, crucial components of the British Indian economy, in which there had been several experiments with malarial prophylaxis to protect some of the most valuable members of the Indian labour force. Yet the report had highlighted just how useless such schemes were if the quinine used was adulterated and had concluded that in such schemes there were 'even greater possibilities of fraud and error' than in hospital dispensaries.

Such a threat to the British economy in India resulted in the establishment of the Drugs Advisory Committee chaired by R.N. Chopra which conducted the first India-wide survey of medicinal drugs with witnesses called and samples taken from a wide range of government and private suppliers. Even so, the commitment of the Government of India to dealing with the problem of drug adulteration could be questioned almost immediately. The Committee had a wide remit for what was essentially a short-term tour. The first witnesses were heard on October 17, 1930, and Chopra's report was presented on March 29 the following year. The tasks of the Committee during this period included ascertaining the level of adulteration of drugs in the medicinal market, recommending methods of controlling both the drug trade in India and the newly emerging pharmaceutical industry, considering legislation to control the trade in indigenous drugs and to create a fully qualified pharmacy profession in the subcontinent. The speed with which the report was finalised, though, could also be put down to the determination of Chopra to deal with the problem of drug adulteration.

The report found 'a very large percentage [of the samples provided] were either adulterated or contained comparatively smaller proportions of quinine than that claimed by the manufacturers'. The Committee repeated Megaw's call for periodic inspections and also recommended improved training, qualifications and pay structure for compounders. It also recommended an overhaul of the Indian Penal Code since there was no specific offence against adulteration of drugs. Finally it proposed the creation of a central specialist laboratory service arguing that it was beyond the capacity of smaller units to analyse the standard of prescription drugs.[30] Few of the recommendations were acted upon immediately, with both central and local authorities pleading financial stringency arising from the depression. Some Provincial Governments looked to be acting, but mostly they were adding a few clauses on drugs to already weak legislation against the adulteration of food.[31] However, a Biomedical Standardisation Laboratory was eventually established in the Calcutta Medical School under Chopra. During 1937 and 1938 his team analysed over 150 samples of quinine and febrifuge products from government hospitals and dispensaries throughout India. The products tested ranged from stock quinine sulphate mixtures to tablets, powders and tinctures of quinine sulphate and quinine bihydrochloride, as well as the tannate tablets used in the treatment of children. Some manufacturers also provided samples direct from the factory. All samples were sealed for transport to Chopra's laboratory. Cinchona samples were tested against the Totaquina standard of the British Pharmacopoeia of 1932 and the British Pharmaceutical Codex of 1934.

Of the 100 samples of stock quinine solutions provided by hospital and dispensary compounders, only 20 were found to contain the stated amount of alkaloid: 64 were under-strength. Twenty-three samples contained less than 60 per cent of the stated dose, 11 of which ranged from no alkaloid to less than 30 per cent. Chopra and his colleagues found this 'a very disquieting state of affairs', since most of these samples came from government hospitals and dispensaries, and they were afraid that results would have been even poorer if samples from private and charitable dispensaries had been included. They argued that 'slight variations', including 16 samples which were over-strength, could be overlooked as simple inaccuracy or 'gross neglect'. However, when they found samples that contained 80 per cent or less of the stated quinine content, they concluded that 'there is strong reason to believe that some fraud is being perpetrated in the compounding room or at the dispensing counter and the situation demands prompt remedy'.[32]

When they analysed 20 samples of tablets, they discovered that 50 per cent were under-strength, including one sample with no alkaloid content at all: place of manufacture made no difference to the results, with 50 per cent of both Indian- and foreign-made tablets being under-strength. However, they consoled themselves with the thought that most tablets had nearly 90 per cent of the stated content and therefore raised no serious therapeutic considerations. Of 28 samples of quinine powders, 13 were under-strength including all samples of cinchona febrifuge and a preparation called 'ionized quinine', which proved to have only 'negligible traces of quinine'. This kind of product angered them, since 'the name leads the public to believe that the preparation contains at least a certain amount of quinine . . . and creates a false sense of security in the minds of the consumers, and in the interests of the public their sales should be banned'.[33] Adulterants included chalk, flour, salt and bicarbonate of soda. Chopra and his co-workers concluded that,

> The results indicate in no uncertain terms that the unsatisfactory state of affairs which existed 30 years ago in this country with regard to the deficiencies in the strength of quinine mixtures and other preparations of similar category and which and been the subject of repeated warnings, still continues with unabated intensity even to this day. If the condition has changed at all, it has changed for the worse . . . The sale of poor quality quinine preparations in India or any

> tampering with the quality of or strength of quinine prescriptions that may take place in the hospitals or dispensaries ... constitutes one of the most cruel forms of injury to the public.[34]

The report led to a strongly worded editorial in the *Indian Medical Gazette* which was quick to lay blame not only on those involved in adulterating the drugs, but also on the colonial authorities who had pled financial restraint as an excuse for inactivity after 30 years of warnings. 'It is hard to believe that those who refuse to take any measures to remedy such a state of affairs and who repeatedly shelve legislation directed towards the proper control of the drugs trade can appreciate the significance of some of these findings'.[35] It was now evident that the root of many of the failures of quinine attributed in the past to the prejudice and ignorance of the patient had a more sinister origin. Bose and his colleagues declared that,

> many of the fallacies in the field of quinine treatment of malaria may be explained by the fact that quinine administered in proper dosage was not possible in a large percentage of cases because of the distribution in the hospital and dispensary of under strength quinine mixtures and of the sale in the Indian market of quinine tablets and quinine salts of inferior quality.[36]

The origins of the 'quinine fraud'

Chopra's findings are important as they suggest reasons both for the failure of much cinchona-based treatments and for the resulting controversies about the efficacy of those treatments. After all, those treatments had remained controversial throughout the nineteenth and twentieth centuries because there remained considerable experience of their failure in medical practice. In 1884 Alexander Crombie lamented to the First Indian Medical Congress that 'there are many practitioners who, from timidity or wrong teaching or some other cause, have no knowledge of, and I feel no belief in, the power of quinine as an antidote to malarial poisoning'.[37] The evidence for such despair was provided by a paper presented to the conference's Pharmacological Section in which Sir William Moore was emphatic in his opinion that 'quinine is a much over-rated drug'.[38] Fifteen years later, Leslie, the Sanitary Commissioner for the Government of India, argued that such a pessimistic view was still widely held with possible just cause:

> This doubt is in most cases founded upon experience – medical men have found persons who have apparently been regularly taking quinine get fever, and most practitioners have come across cases in which quinine seemed to fail to cure malarial fever, while instances are not uncommon in which the results of malaria prophylaxis in a community have been exceedingly unsatisfactory.[39]

As late as the 1930s one eminent specialist in tropical medicine reported that 'there has, almost from its first introduction, been a dispute concerning the giving of quinine, the dosage, the intervals between doses, the time relative to fever, whether it should be given alone or with other drugs and so forth. The question can hardly be regarded as really settled even today'.[40] Throughout these controversies, proponents of quinine therapy blamed problems on patient failure to follow instructions rather than the drug itself.

In pointing the finger at 'those who refuse to take any measures to remedy such a state of affairs and who repeatedly shelve legislation', Chopra and his colleagues correctly identified the colonial authorities as to blame for this state of affairs. In the first place, the team reiterated the findings of Megaw in 1907 that one of the roots of the problem lay in the low status of compounders in India which resulted from poor training and low levels of pay. Few compounders had more than a basic apprenticeship as training. Courses had been established in Madras in 1899, in the United Provinces in 1928 and in Bengal in the following year. However, few took these courses and there was a major shortage of compounders as only 7102 worked on the entire subcontinent in 1937.[41] This was a problem that was consolidated by the limitations of the Indian Medical Service, whose members were few in number and grossly overworked with little time to check the efforts of the compounders.

Moreover, there was growing confusion about the role to be played in public health by compounders. For instance, James' experimental camps would initially be set up by a Medical Officer, but would then be left in the hands of a compounder.[42] There was a growing tide of complaint in the inter-war period about compounders effectively acting as untrained doctors, yet the authorities had colluded in the creation of this situation. At a time of financial stringency, the use of compounders as 'medical assistants' proved to be a cheap contingency which few local authorities could overlook. In the United Provinces, compounders 'trained by Civil Surgeons in plague inoculation work and in the treatment of plague' during a major epidemic soon found their duties extended. Colonel MacTaggart, the Inspector-General, declared, 'they have also been utilised in connection with severe epidemics of cholera, relapsing fever, malaria and influenza. They have been very useful also at fairs and melas . . . The increasing numbers of patients who resort to them for medical and *surgical* relief bears testimony to their usefulness and popularity'.[43] With a seven-point pay scale for male compounders ranging from Rs. 18 to Rs. 50 per month and for women from Rs. 10 to Rs. 30, compounders provided a cheap option who were being given ever greater responsibilities.[44]

Western pharmacy was slow to develop in India, having a low standing in government priority lists. A post in Materia Medica was created in the United Provinces Medical Hospital but it was combined with forensic medicine.[45] It was not until 1934 that M.L. Shroff had instituted a Pharmaceutical Chemistry course at Benares Hindu University, followed by a full Pharmacy course 2 years later. As Sanyal argued, 'Prior to independence there was no corresponding growth in qualified pharmacists commensurate with the increase in numbers of physicians'.[46] The report of the Biomedical Standardisation Laboratory echoed earlier calls for increased testing of dispensaries and spot checks in the bazaar, and Chopra was concerned that even if Medical Officers were enthusiastic about testing, few had the training in biochemistry and pharmacology to conduct more than the most basic level of analysis. Chopra went on to recommend an expansion of laboratory facilities at both central and provincial level and training at university and college level to increase the standing of pharmacy as a profession in India. This would naturally lead to increased levels of pay, he believed, and a reduction in the temptation to fraudulent practices.

In colonial South Asia, only a small cadre of officials were trained sufficiently in chemical and pharmaceutical sciences to test medicinal substances for adulteration: the members of the Chemical Examiners' Department. As was occurring in metropolitan Britain, the role of the Chemical Examiner in pre-independence India was growing ever more complex over the last 60 years of British rule. However, whereas in the United Kingdom this increasingly resulted in the emergence of specialisations within the profession, in India the various duties generally were carried out by a small staff under a Chief Chemical Examiner who was expected to be competent in all the required areas. As well as the multiple roles of chemical analyser, including the evaluation of spirit level for excise and customs duty purposes and quality testing water, food and drugs, they provided medico-legal reports for the local Police and acted as Professors of Chemistry in local Medical Colleges. If this was not difficult enough, not every Local Government could afford to employ a Chemical Analyst Department. So, for instance, the Bengal Examiner also tested samples from Assam, Bihar and Orissa, while the Examiner to the Government of the Punjab automatically assumed the role for the Government of the North-West Frontier Provinces as well. In their chemical analyses, however, checking for the purity of medicinal drugs was a low priority. Instead, by the turn of the century, they were literally deluged with narcotics seized by the Excisemen. This contraband had to be examined in order to assess if it was fit for human purpose, thus allowing the Excise Department to sell it on to the Medical Stores Depots. Windsor, in Burma, complained that 'it is required that every bottle and every packet must be individually tested': no mean feat when by 1910 his laboratory was testing 14,950 samples of cocaine and 4358 of morphia. This contrasted to the 18 quinine tablets tested in the same year.[47] While this was the laboratory most involved in narcotic testing, Hankin in the United Provinces tested 1306 samples of cocaine against two quinine tablets in 1914.[48] Thus the safeguarding of supplies of medicinal drugs was subordinated to the needs of financing the Raj on the one hand and sublimated to the obsession with narcotics on the other.

Chopra also argued that another factor in the problem was the very success in inculcating a demand for Western medicine in India. Since few could afford such products, 'unscrupulous manufacturers', both domestic and foreign, were abusing lax legal definitions of drugs to 'flood the market with an inferior grade of product'. The 'quinine fraud' was only one example replicated in most pharmaceutical products tested by the Biomedical Standardisation Laboratory. Chopra demanded stronger legislation of the drug market and pharmaceutical industry in India to prevent adulteration with spot testing at every stage from manufacture or import to sale in the bazaar or dispensing in hospitals or clinics.[49] Such legislation would be vital since there was little in the Indian Penal Code which prevented the adulteration of medicinal drugs. Police had to prove criminal intent to harm which was difficult when few standards had been set for the composition of drugs. The few convictions that were successful had come under the terms of the Indian Merchandise Marks Act which was largely utilised against false claims of patent medicines. Meanwhile few provincial governments had legislation enforcing the training of compounders and the licensing of drug retailers, and those that did, such as Bengal, rarely enforced the legal requirements.

Yet there were other problems in colonial policy on cinchona products that served to further complicate the picture. The most obvious is the failure of the self-sufficiency policy which meant that, with a limited supply and an extensive demand, there was the scope for private profit to be had from trading in the products. This failure came despite renewed efforts to achieve self-sufficiency in the twentieth century. As a result of the 1908 epidemic which devastated the United Provinces, the Government of India decided to create a quinine reserve for use in any future crisis and to guard the domestic price against inflation in world prices. Domestic production was to be supplemented by imports from the Dutch East Indies until such times as local production increased. The Great War once again boosted the ideal of self-sufficiency. However, production costs in India remained high. Large quantities of bark were required for the extraction of the alkaloids: in Bengal from 1.4 million tons of bark, 51,094 tons of quinine was extracted and 28,523 tons of cinchona febrifuge.[50] With a 10- to 15-year time lag before the bark was ready for harvest, future demand was difficult to estimate and few private entrepreneurs were willing to take the risks involved in production. Therefore, it was left to the Indian authorities to subsidise the plantations.

The collapse of the price of silver as the war drew to a close left the Government of India with a post-war legacy of financial stringency. Public health did not escape the Inchcape Retrenchment Committee which advised the government to cut back on services until India could meet its financial obligations to Britain. However, while abandoning the policy of self-sufficiency again, the Government of India continued to uphold the quinine emergency reserve policy. Ironically, given the cost of quinine, it was still viewed as a cheaper treatment than the types of sanitary anti-malarial campaigns that involved draining land, lowering groundwater levels and clearing jungle, measures that had been championed by Ronald Ross. The inter-war era would see the replication of the often bitter debates between the protagonists of quinine therapy and sanitary measures that had developed during the late nineteenth century.[51] The Inchcape Committee then recommended that the government reserve should be capped at the existing figure of 217,000 lbs, and that no more quinine should be purchased.[52] The search for cheaper methods of mass treatment in non-epidemic seasons, therefore, was now a matter of urgency.

By the end of the decade, however, the Royal Commission on Agriculture in India was signalling another u-turn in official policy. It declared that 'if India is to embark on any large campaign of fighting malaria, we are convinced that it will be necessary to reduce considerably the price of quinine within India and this can only be effected if India is self-supporting in production. To achieve this self-sufficiency, a considerable extension of cinchona will be required'.[53] Again Government support would be required during the long cultivation period, particularly given the ever-present possibility of climatic calamity in India. However, the Report was presented just as the inter-war depression once more made government finances precarious and it was not until the late 1930s that the scheme would be revived under the renewed shadow of war.

Cost also drove the distribution strategies of the colonial state. It was accepted practice during a malaria epidemic for local authorities to provide quinine free of charge, as occurred in the Punjab and the United Provinces in the epidemic year of 1908. In the former, some 6000 lbs was distributed by the local authorities.[54] There were also some experiments in providing quinine or cinchona febrifuge free as a prophylactic. However, in 'normal' years, authorities usually preferred to sell quinine through a complex network of distributors. This was for a number of reasons. Some claimed that the free supplies were simply thrown away, stored up until useless or had 'actually been sold in the bazaar for what it would fetch'.[55] Others thought that the free supplies were not used properly; once the first phase of fever ended so did the consumption of quinine, which ensured that the patient still had parasites in their body which guaranteed a recurrence of the problem. The solution to this, it was argued, was to make the patient pay for the treatment in which case the individual would take the full course in order to get their money's worth.

Once the decision was made to reduce the free distribution of quinine as much as possible, authorities began to experiment with an ever-widening network of salesmen. Initially sales were confined to Post Offices and government officials from the Public Health and Excise and Stamp Departments. While some thought this provided a sufficient and well-known distribution network, others raised doubts whether such vendors had been advertised sufficiently. In 1901 James argued in opposition to Ferand's claim that villagers in the United Provinces knew where they could get supplies of quinine, that 'It is certain at present... not a fractional proportion of [villagers in India] know where the medicine is to be obtained.'[56] Such worries ensured that more groups were engaged as vendors. 'We find that postmasters are our most successful salesman [but] we cast as wide a net as possible and employ schoolmasters, members of the village panchayats, chaukidars and shopkeepers'.[57] In the early stages, it was clear that the authorities were trying to establish an official network. James insisted that Post Offices should put up a sign stating 'Government Quinine sold here', while Ferand declared that the vendors chosen were 'the most trustworthy officials of various government departments'.[58] However, over time networks blossomed into a highly complex web of authorised distributors, including vaccinators and plague staff, itinerant dispensers, missionaries, schoolteachers, village headsmen and tax and rent collectors. In the Punjab, licences were supplied to most of these categories and, as one official put it, 'any other respectable persons recommended for the work'.[59] By 1937, supplies of quinine and cinchona febrifuge were being sold from 398 government hospitals and dispensaries, supplemented by 670 'special centres' and 3595 sub-centres in what was regarded as 'a relatively quiet year' for malaria.[60] Zamindari agents were utilised in the United Provinces. Anti-Malarial Co-operative Societies and Village Uplift and Rural Reconstruction Schemes also became part of the distribution network. The sheer complexity of the sales distribution precluded organised control by government authorities.

The most important element of the commercialisation of the distribution network, and the possible motive for adulteration of stocks, was that these agents were unpaid. They bought the quinine and febrifuge products from the local government, district or municipal authority and sold it on to the public for a profit. Attempts to create large networks of voluntary groups failed and so the possibility of financial gain was used as the incentive to distribute the drugs. Before 1914, profits ranged from 1.5 annas per box of 10 packets in the Central Provinces, where sales were said to be small with little incentive to the vendors, to 9 annas in Bengal, a commission rate of 33 per cent.[61] Major Wilkinson of the Eastern Bengal IMS explained the policy: 'This may appear high but it is not, I submit, excessive. Relying as we do on the services of an unpaid agency, it is essential that sales are pushed automatically, and it seems doubtful whether any scheme of distribution which does not satisfy this condition is likely to succeed'.[62] Gill accepted that the networks had only operated successfully in the Punjab once the commission had been raised from 2 to 4 annas per treatment in a deliberate attempt to 'increase the interest of postmasters in the sale of quinine and to stimulate them into greater activity'.[63] Since inspections of the distributors was rare, it was likely that prices and profits were far higher than the 'official' limits in such an uncontrolled trade in an expensive therapy.

Evidence for this can be gleaned from the failure of an imaginative pre-1914 scheme in Assam to transform vendors into official state employees. Three full-time vendors, 'specifically selected by the Civil Surgeon', were to be appointed on a subsistence allowance of Rs. 10 per month plus a commission of 7 annas for every quinine parcel sold.

Vendors would not have to purchase quinine as the local government would supply them with treatment packets and leaflets in the vernacular stating 'quinine cures malarial fever but no other fever. It can be bought from the vendor or from Post Offices for 3 annas per tube.'[64] Within 12 months, however, the Lieutenant-Colonel Banatvala, the Sanitary Commissioner, was reporting failure 'as the right man for the job could not be obtained for the post'.[65]

The official position was summed up by James, one of the foremost malariologists: 'In the attempt to mitigate malaria in India, the Government and local authorities have an obvious duty to perform, but is not limitless'.[66] Instead the object of selling cinchona products was to inculcate the idea of 'self-treatment' of malaria. While loss of revenue was accepted as part of the policy of selling quinine and febrifuge, it was still cheaper than the cost of undertaking widespread free distribution. To publicise the new policy a wide range of advertising was utilised including leaflets in the vernacular, exhibitions, travelling film shows, pictures of mosquitoes on boxes of tablets, and printing messages on the front covers of schoolbooks.[67] As Wilkinson explained, 'if we are to oust quack medicines and fever mixtures we must fight them on their own ground and appeal to the masses in the manner calculated to impress them'.[68] Even children in the Scouts and Junior Red Cross Brigades were pressed into service providing positive propaganda for the sales.[69] This created a powerful contradiction: 'Propaganda, although ordinarily of value, must largely fail if the advertised drug is beyond the villagers' financial resources. The question of price is therefore intimately associated with that of propaganda campaigns and the two cannot be considered separately.'[70] Yet by utilising the profit motive in the distribution network, the government was colluding in pushing up the price of the treatment.

Conclusion

It was Government collusion in a number of processes that encouraged the 'fraud... perpetrated in the compounding room or at the dispensing counter' exposed by Chopra's team. While this must be set in the context of the limited reach of Western medicine in the subcontinent, with estimates that quinine and febrifuge consumption ranged from only 1.1 to 8 grains per capita per annum, for those using the adulterated products there could be serious consequences. The situation arose from a combination of low pay and lack of status for compounders, dispensers and hospital assistants, an overly complex network of vendors selling quinine for a profit and a lax legal definition of what constituted a drug offence beyond the realm of narcotics control. The common theme in examining all of these factors was money. The colonial state poorly paid compounders, dispensers and hospital assistants in order to limit the cost of its healthcare system and chose to turn to a network of local vendors as it seemed cheaper to do so. It rarely employed its Chemical Examiners or other scientists on the business of quality control of medicines as it focused their efforts on the more lucrative activity of guaranteeing high-grade narcotics reached the markets for opium and cannabis.

Despite its antiquity, during this period quinine continued to be regarded as an experimental anti-malarial, one which by 1939 was considered to have failed. Medical professionals regularly reported cases of failed quinine administration in hospitals, while few believed in its power as a prophylactic by the onset of the Second World War. Doctors eagerly awaited the cheapening of the new laboratory synthesised anti-malarials such as plasmoquine so that they could replace quinine in general practice. It seems clear, though, that a large part of its flawed reputation was based on a serious misuse of the drug. Between 25 and 50 per cent of the quinine and febrifuge products that were in circulation in colonial South Asia were under the strength of their stated contents. This inadequate and inconsistent dosage resulted in a reduced curative and prophylactic effect. It deepened suspicions of Western medicine at a time of rising nationalism in the subcontinent. It compromised the effectiveness of government quininisation campaigns, raising more doubts about the British ability to improve the lot of colonial populations. Adulteration of quinine undoubtedly aided the build-up of resistance to the drug and caused unpleasant side effects for those taking the under-strength preparations. Above all, adulterated quinine failed to cure those using it as an anti-malarial therapy. The death toll from such actions within the wider pattern of mortality from malaria in British India can only be a matter of conjecture.

Notes

1. Editorial, 'The Quinine Fraud' in *Indian Medical Gazette* [hereafter *IMG*], 74 (October 1939), p. 623.
2. Annual Sanitary Report of the Province of Assam for the Year 1902 (Shillong: Assam Secretariat Printing Office 1903), p. 3.
3. L. Rogers, 'Gleanings from the Calcutta Post Mortem Records' in *Glasgow Medical Journal*, 1925.
4. Sir Ronald Ross, *Memoirs with a Full Account of the Great Malaria Problem and Its Solution* (London: John Murray 1923).

5. H.W. Acton and R. Knowles, 'On a Standard Treatment for Malaria' in *IMG*, 59 (April 1924), p. 198.
6. Lord Minto, 'Opening Speech', First Imperial Malaria Conference, 1909, Cd. 6538 of 1912, p. 5.
7. A.J.H. Russell, 'Quinine Supplies in India' in *Records of the Malaria Survey of India* [hereafter *Records*], 8, 1937, p. 238.
8. J. Sinton, 'Human Costs of Malaria in India', *Records*, 7, 1936.
9. C.A. Gill, 'Summary of Anti-Malarial Measures in the Punjab', in Cd 6538 of 1912, p. 61.
10. S.R. Christophers, 'On Malaria in the Punjab', 1909, Cd 6538 of 1912, p. 41.
11. D. Arnold, *Colonizing the Body. State Medicine and Epidemic Disease in Nineteenth Century India* (Berkeley: University of California Press 1993), p. 202.
12. M. Harrison, 'Hot Beds of Disease': Malaria and Civilization in Nineteenth Century India' in *Parassitologia*, 40, 1998, p. 11.
13. F.F. Cartwright, *Disease in History* (London: Hart Davis Ltd 1972), p. 150.
14. E.H. Ackerknecht, *Therapeutics: From the Primitives to the 20th Century* (New York: Hafner Press 1973), p. 115.
15. Russell, 'Quinine Supplies', p. 235.
16. Acton and Knowles, 'On a Standard Treatment for Malaria'.
17. Russell, 'Quinine Supplies', p. 238; Annual Report of the Public Health Commissioner to the Government of India for the year 1938 (New Delhi: Government Press 1940), p. 48.
18. Annual Report of the Cinchona Plantations and Factory of the Government of Bengal for the Year 1930–31 (Calcutta: Government Publishing, 1932), p. 5.
19. Annual Report, Public Health Commissioner to the Government of India, 1933, volume I (New Delhi: Government Press 1935), p. 3.
20. G. King, 'The Government Cinchona Enterprise in Sikkim', First Medical Congress, p. 53. [Italicized in the *Proceedings*.]
21. Ibid., p. 530.
22. Leslie, 'Malaria in India', in Cd 6538 of 1912, p. 12.
23. S.R. Christophers, 'On Malaria in the Punjab', Cd 6538 of 1912, p. 41.
24. J.D.W. Megaw, 'A Year's Experience of Malaria at the Outdoor Department of the Medical College Hospital, Calcutta', *IMG*, 42 (January 1907), p. 13.
25. Gill, 'Anti-Malarial Activities in the Punjab', 1909, Cd 6538 of 1912, p. 61.
26. Sanitary Report for Burma for the Year 1905 (Rangoon: Government House 1907), p. 29.
27. Discussion, 1909, Cd 6538 of 1912, p. 78.
28. November 11–16, 1912, Conference Proceedings, volume II (Simla: Government Central Branch 1913), p. 530.
29. J.W.D. Megaw, S. Ghosh and N.R. Chatterjee, 'Stock Solutions of Quinine', *IMG*, 63 (May 1928), pp. 244–247.
30. R.N. Chopra, Report of the Drugs Advisory Committee, 1930–1931 (Delhi: Government Printing Press 1932).
31. See, for instance, Bihar and Orissa and the United Provinces in 1933, Report of the Public Health Commissioner of the Government of India in 1933, p. 279.
32. I.Bose, B. Mukerji and R.N. Chopra, 'Quality of Quinine Preparations in Indian Hospitals and Dispensaries', *IMG*, 74 (October 1939), pp. 610–1.

33. Bose *et al.*, 'Quality of Quinine', p. 612.
34. Ibid., pp. 610 and 612.
35. Editorial, 'The Quinine Fraud', p. 623.
36. Bose *et al.*, 'Quality of Quinine', p. 612.
37. Presidential Address, 'On the Fevers of India', First Indian Medical Congress, *Proceedings*, January 1894, p. 19.
38. Sir W. Moore, First Indian Medical Congress, *Proceedings*, p. 531.
39. Leslie, 'Malaria in India', Cd 6538 of 1912, p. 12.
40. H.H. Scott, *A History of Tropical Medicine* (London: Edward Arnold & Co 1939), p. 232.
41. P.K. Sanyal, *A Story of Medicine and Pharmacy. Indian Pharmacy 2000 Years ago and After* (Calcutta: Amitava Sanyal 1964), pp. 142–4; Report on the Civil Hospitals and Dispensaries in the United Provinces for the years ending 1927–1930.
42. S.P. James, 'Experimental Demonstration Camps' 1909, Cd 6538 of 1912, p. 76.
43. Annual Report, Civil Hospitals and Dispensaries, United Provinces, 1918, pp. 4–5. My italics.
44. Triennial Report, Civil Hospitals and Dispensaries, United Provinces, 1919–1922, p. 9 and Annual Report for 1918, p. 11.
45. Annual Report for 1918, p. 7.
46. Sanyal, *Story of Medicine and Pharmacy*, p. 179.
47. Chemical Examiner, Burma, 1908, p. 1 para 2; 1910, p. 1 para 2.
48. Chemical Examiner, United Provinces, 1914, p. 2, para 10 and Table VI.
49. R.N. Chopra, 'The Drug Industry in India and its Difficulties', speech reprinted in *IMG*, 74 (April 1939), pp. 230–1.
50. Annual Report, Public Health Commissioner for the Government of India, 1933, p. 66.
51. For the nineteenth-century debates see Rogers, *Happy Toil* and for the post-1918 period see *IMG*, 65 (December 1930), 'Malaria Control in India', which discussed the increasingly heated correspondence on the issue in the *British Medical Journal* between March and September 1930 involving leading malariologists including C.A. Gill, Sir Ronald Ross and Sir Malcolm Watson.
52. Russell, 'Quinine Supplies', p. 236.
53. Royal Commission on Agriculture in India, 1929.
54. Gill, 'Anti-Malarial Measures in the Punjab', Cd 6538 of 1912, p. 60.
55. Surgeon-General Bannerman, 'Anti-Malarial Measures – Itinerant Dispensaries', Second All-India Sanitary Conference, Madras, 1912, *Proceedings*, Cd 6777 of 1913, p. 18; Wadhia in discussion, First Malaria Congress, Cd 6538 of 1912, p. 84.
56. James, 'Experimental Demonstration Camps',1909, Cd 6538 of 1912, p. 76; Ferand, p. 78.
57. James, p. 75.
58. James, p. 77; Ferand, 'Discussion', p. 78.
59. Gill, 'Anti-Malarial Measures in the Punjab', 1909, Cd 6538 of 1912, p. 61.
60. Annual Report of the Public Health Commissioner for the Government of India, 1937 (New Delhi: Government Press, 1919), p. 46.
61. Central Provinces delegate, Fox-Strangeways, Discussion, Imperial Malaria Conference, Cd 6538 of 1912, p. 84.

62. J. Wilkinson, 'Revised Scheme for the Distribution of Quinine by Government', Cd 6538 of 1912, p. 74.
63. Gill, 'Anti-malarial Activities in the Punjab', 1909, Cd 6538 of 1912, p. 59.
64. Annual Sanitary Report of the Province of Assam for the Year 1913 (Shillong: Assam Secretariat Printing Office 1914), p. 21.
65. Annual Sanitary Report, Assam, for the Year 1914 (1915), p. 20.
66. James, 'Experimental Demonstration Camps', 1909, Cd 6538 of 1912, p. 78.
67. Gill, 'Anti-Malarial Activities in the Punjab', 1909, Cd 6538 of 1912, p. 62.
68. Wilkinson, 'Revised Scheme', 1909, Cd 6538 of 1912, p. 75.
69. Annual Report of the Public Health Commissioner for the Government of India, 1935, p. 64.
70. Public Health Commissioner for the Government of India, 1935, p. 41.

Part III

'High' Politics

10

Colonial Africa and the International Politics of Cannabis: Egypt, South Africa and the Origins of Global Control

James H. Mills

Introduction

In exploring the origins of international regulations governing movements in such intoxicating drugs as opium and cocaine, historians have pointed to the jostling of Western empires in Asia.[1] Indeed, elsewhere in this book, William McAllister and William Walker return to the rivalries of the British, the Americans, the Chinese and the Japanese in shaping the international agenda in the first decades of the twentieth century that drove the development of treaties and laws that grew out of the Opium Conferences. This chapter reconsiders the opium politics of the 1920s in order to explore the origins of controls on cannabis. It argues that in considering the sudden appearance of cannabis as an issue in the 1920s at the League of Nations, it is necessary to turn to colonial Africa, rather than to Asia, to seek explanations.

Cannabis had briefly featured in discussions at The Hague in 1912, where it was introduced by the Italians. However, it did not seem clear even to that country's representatives why they did this, and it was noted that 'the Italian delegation was not to present any specific proposal in regard to hemp drugs, but rather to leave it to the conference as a whole to take such measures in regard to these as it might deem expedient'.[2] Needless to say, no mention of cannabis made it into the International Opium Convention signed at The Hague on January 23 1912. This focused entirely on 'the gradual suppression of the abuse of opium, morphine and cocaine' and committed the contracting countries to a range of laws and regulations controlling the manufacture, sale and use of those drugs in their various forms.[3]

Cannabis was not to feature again in the international discussions until 1923. By then the meetings had been formalised under the aegis of the League of Nations which had spawned an Advisory Committee on Traffic in Opium. This had been meeting regularly since 1921. By the end of the Fifth Session of the Advisory Committee in 1923, much heated debate had been generated but little progress had been made on the key issue of opium production. Cannabis meanwhile had been altogether forgotten by those on the Committee. However, this was about to change. Towards the end of 1923 and while the Advisory Committee was in recess, a letter arrived in The Hague for the consideration of the group when it reconvened in 1924. It read as follows.

> Pretoria November 28th 1923
>
> With reference to your letter no. 12/A/22951/17217 dated September 6th 1922, on the above subject and to my letter no. 29/8/85 dated December last, forwarding copies of the Regulations promulgated under Proclamation no. 181 of 1922, I have the honour to inform you that, from the point of view of the Union of South Africa, the most important of all the habit-forming drugs is Indian hemp or 'Dagga' and this drug is not included in the International List. It is suggested that the various Governments being parties to the International Opium Convention should be asked to include in their lists of habit-forming drugs the following:
>
> Indian hemp: including the whole or any portion of the plants cannabis indica or cannabis sativa.
>
> Signed, J.C. Van Tyen, for Secretary to the Prime Minister.[4]

The letter was not considered until August 12 1924, at the Twelfth Meeting of the Sixth Session of the Advisory Committee on Traffic in Opium and Other Dangerous Drugs. The British delegate Malcolm Delevingne and John Campbell of the India Office made reference to the letter from South Africa in order to draw attention to the cannabis issue. It was Campbell who piped up first and he assured the Committee that the Government of India was already dealing with the issue. He told his colleagues that it had been informed of the South African communication and that it had acted promptly to consult the various administrations in India so as to discover whether it would be possible to prohibit the export of Indian hemp. He suggested leaving a full consideration of the subject for future sessions in order to allow time for information to be collected. Delevingne supported this suggestion and pointed out that the British government was also well ahead of others as it too had been collecting information on the subject of Indian hemp. Indeed, he even had a report to hand about the extent of the problem that the British and the Government of India had made clear they were already doing so much to investigate.

> Ten tons of this plant had recently been exported to Abyssinia, the owner of the consignment alleging that it was to be used as an insecticide. The consignment had reached the Port of Djibouti but eventually after a series of operations, it had disappeared entirely. Enquiries made by the British and French Governments left little room for doubt that it had been smuggled into Egypt for illicit purposes.[5]

The American delegate, Edwin Neville, seemed unconcerned about the issue and asked pointedly if it was the case that any Government had actually expressed a desire for the question to be discussed. Campbell was quick to reply, although not directly to his question, stating that 'the Government of India had taken up the question in advance from the point of view of prohibiting export should that appear desirable'. The Committee approved Malcolm Delevingne's proposal that Governments should be asked to provide more information on Indian hemp and a questionnaire was circulated.

The matter might have met the same fate as the Italian interest in the same subject of 1912 had it not been for the Second Opium Conference that was assembled in 1924. While the Advisory Committee only invited delegates that claimed to have direct interests in the opium issue, the Conference was open to representatives of all of the countries that were members of the League of Nations. It was assembled solely to meet two specific objectives. The first was to devise schemes to set maximum limits on the amounts of morphine, heroin and cocaine that could be manufactured from raw opium and cocoa and also to restrict the production for export of the raw materials from which those drugs were made. The second was the amendment of the Hague Convention of 1912 so as to take account of the schemes devised to achieve the first objective.

Dr Mohamed A.S. El Guindy, the head of the Egyptian delegation, was determined to take the agenda elsewhere, however.

> The illicit use of opium and its derivatives and of the other substances mentioned in the Advisory Committee's report is universally condemned by public opinion. There is, however, another product which is at least as harmful as opium, if not more so, and which my Government would be glad to see included in the same category as the other narcotics already mentioned – I refer to hashish,

> the product of the cannabis indica or sativa. This substance and its derivatives work such havoc that the Egyptian Government has for a long time past prohibited their introduction into the country (except of course the trifling quantity required for medical purposes). I cannot sufficiently emphasise the importance of including this product in the list of narcotics the use of which is to be regarded by this Conference. I hope soon to be in a position to submit to this Conference a short memorandum on this question which is of such importance to my country.[6]

From this opening gambit, El Guindy successfully forced cannabis into the deliberations of the meeting and finally into the agreements made there. He did this despite the obstacles presented by the British, by cleverly aligning himself with powerful allies such as the United States, who were at loggerheads with the latter on opium matters.[7] The rest of this chapter will begin to explore the reasons why it was governments from opposite ends of the African continent who were so concerned to push cannabis substances into the emerging systems of international drugs control.

Dagga, labour and commerce in colonial Natal

Between 1834 and 1938, 5.5 million Indians migrated as labourers. The first workers went to Mauritius from Madras and Calcutta in 1829, and by the 1870s, Indians were regularly going to 12 colonies within the Empire, including Natal, Jamaica, Trinidad, British Guiana, Fiji and also to French Réunion. These migrations were deliberately stimulated by the British authorities in these places as well as in India itself in order to meet the labour demands of the imperial economy.[8]

That Indians took this fondness for cannabis substances with them when they migrated to work within the Empire is most clearly demonstrated in the case of Natal. The Indian Immigrants Commission of 1885–1887 conducted an extensive survey of the extent to which the Indian population of the southern African colony used cannabis. It concluded

> a. that the smoking of hemp, whether by itself or in the mixture to which we referred, is detrimental to the health of Indian Immigrants in this Colony,
> b. that the immoderate use of it is highly injurious

c. that the habit of smoking it in excess is widespread, in the Pietermaritzburg Circle one-fifth probably, of the Indian population smoking it in excess
d. that such immoderate use leads to crime of the most serious nature
e. that it renders the Indian Immigrant unfit and unable to perform, with satisfaction to the employer, that work for which he was specially brought to the Colony.[9]

The Commission's conclusions were based on some almost hysterical reporting of cannabis use and its effects among the Indian workforce. Witnesses such as Captain GA Lucas, the Resident Magistrate of Alexandra County, conjured up horrifying stories of violence and madness:

> As to the smoking of dakkha there is at this present minute in the gaol here an Indian, Bhalee by name, an indentured Indian, who several months ago was brought in on a charge of assaulting a woman in the magasse yard of the Nil Desperandum estate (Bazley's); the assault was a serious one and he was sent to gaol for a fortnight. The Indian Medical Officer (Dr Tritton) reported to me that the man was under the influence of dakkha and he was under the impression that the prisoner was a habitual dakkha-smoker. When he first came in he appeared dazed and in Court he broke away from the police and attacked the woman who was giving evidence against him. At the end of his sentence he returned to his estate and in a few days committed a serious assault on the same woman; he was sent back in custody, the District Surgeon, Dr Trevor, having certified that it was not safe to leave him at large and now he is detained in gaol pending instructions. I am decidedly of opinion that the law should prevent the smoking of dakkha by Indian immigrants.[10]

To this image of violence and madness was added illness as a doctor reported that 'probably one fifth of the number of Indians smoke the drug to a detrimental extent, and many ailments are due to the use of it'[11], and the Protector of Immigrants himself confidently stated that 'for a long time past cases have frequently come under my notice of Indians, some of them quite young men, who otherwise should be robust, active and healthy, whose strength and manhood seemed to have been sapped by an over-indulgence in the use of this weed'.[12]

Indeed, the members of the Commission itself had an unwelcome encounter with a cannabis user at first hand:

> We ourselves when visiting an estate in the Umzinto circle whereon Indians were employed, came upon an Indian, an absentee from work, sitting outside a hut, with his dakkha pipe on the ground by his side. He muttered to himself, then yelled, spoke rapidly and incoherently, lapsed into silence. Then yelled again, and it was impossible to make him understand anything. He was, manifestly, in a state of dementia induced by dakkha smoking: he was decidedly dangerous, and the Manager was uncertain how to deal with him; finally the man was left to do as he pleased, the Indians on the estate being afraid to interfere with him, and the Manager knowing that the law provided no punishment for his misconduct.[13]

The way in which these reports present cannabis and cannabis consumption provides an insight into the concerns that underpinned them. The Indians were in the country to work and it seems evident that it was feared that indulgence in an intoxicating substance undermined their potential as labourers. Indeed, the account above does not simply show an anxiety about the man's labour but also a sense that the drug has rendered the man reluctant to or incapable of producing the required deference and obedience when confronted by white officials. In a colony where the population by this period was made up of about 370,000 Africans, 36,000 Europeans and 30,000 Indians, it is not difficult to see why situations where the authority of the white man was fragile were feared.[14] Indeed, such was the determination of the Commission to find against cannabis that dissenting voices such as that of Magistrate Paterson in the Estcourt District were ignored as he wrote that 'I have had a few cases in gaol here of Indians suffering from the effects of dakkha-smoking; but I do not think that the vice exists to such an extent in this district as to require special regulations for its suppression.'[15]

Indeed other reasons may have existed that acted against cannabis in Natal in this period. The first was a well-organised Chamber of Commerce that consisted of white shop-keepers and merchants in the Colony. They were eager to sew up the demand for intoxicants by dealing in alcohol and actively sought to keep competitors from this market.[16] Indeed, in places they were condemned for profiting from selling alcohol to Indians as it was felt that the latter were apt to take to drink and also that they were apt to break the colony's laws by selling alcohol on to the African population.[17]

Complicating the picture of the market for intoxicants, however, was the trade in cannabis. The question remains of where the Indian workers in Natal got hold of their supplies. Part of the answer lay in their own initiative as the Commission reported that

> In Natal hemp is cultivated by Indian Immigrants ... in many parts of the colony it grows wild. Some Immigrants smoke it almost in its green state at the time (May) when it is gathered and when its active principle (cannabin resin) and volatile oil are most powerful and particularly obnoxious to the consumer. Some smoke the dried leaves alone. Other Indians consume a mixture compounded of tobacco, opium, hemp, and brown sugar, and the fumes from this compound are not even passed through water to abstract the volatile oils and cannabin resin.[18]

One other source, however, was the local African population. It seems that the locals had used cannabis for the purpose of intoxication long before the arrival of the Indian workers.[19] Indeed, it seems that the Africans were still using cannabis at the time of the IIC as Dr Brewitt of Estcourt could report that 'amongst the natives of Natal the habit is very prevalent and cannot be too strongly condemned. Since I have been District Surgeon in Weenen County, I have examined three natives reported by their friends to be insane. I have found them, at first, very excitable and wandering in giving answers to questions.'[20] Other witnesses, however, provided evidence that makes the idea unlikely that the Indians simply borrowed local habits rather than bringing their own tastes with them. Dr Kretchmar of the Verulam Circle, for example, noted that 'the Indians pleaded that from youth without excess they had been accustomed to this article for smoking and never carried the indulgence to excess'.[21]

What was clear to the IIC was the fact that Africans traded in cannabis with Indians. Witnesses such as Dr Tritton in the Umzinto Circle reported that 'it is very largely used among the natives, from whom the Indians mostly obtained it'[22] but the IIC stated thought the process even more complex: 'we have reason to think that much hemp is sold to Indians by Kaffirs and storekeepers; we are aware that, in some parts of the Colony, white traders purchase green hemp leaves from Kaffir growers and retail them in a dried state to any customer who applies for them'.[23] Cannabis was likely to have been unpopular with the authorities of the colony not simply because it was wound up in their anxieties about productivity and worker defiance. It was also a rare point of contact between the Asian and the African subjects of their rule and a source of profit for the unscrupulous white traders of the region.

Taken together, however, a picture begins to emerge that provides a context for understanding the appearance and the content of the letter from the Union of South Africa to the League of Nations in 1923. When it spoke of regulations on Indian hemp of 1922, it was in fact speaking about controls that governments in the region had been developing since the 1880s. These governments had been driven by a range of anxieties about their Indian labour force, including concerns about their efficiency as workers, their obedience as subjects and their relationships with African neighbours.

Hashish, prohibition and medicine in colonial Egypt

While it is possible to trace the ways in which El Guindy and the Egyptian delegates managed to foist cannabis on to the agenda of the emerging international drugs regulatory system, the League of Nations records do not explain quite why it was that they were so keen to do this. A general concern with cannabis can be traced in Egyptian government for centuries, but El Guindy's sudden assertiveness about the subject and indeed the solutions to the problem that he described so vividly seem to have had rather more contemporary origins. Indeed, it may well be the case that while the British colonial officials of the Government of India were vigorously opposing his assertions, other branches of Britain's imperial administration were unwittingly supplying him both with his evidence and with his arguments.

Cannabis had been widely used in Egyptian society since medieval times and European observers noted that preparations of the hemp plant were still common in the nineteenth century. Lay medication was popular in Egypt where it was commented that 'it is worth remarking about the acquaintance the inhabitants of Egypt have with a great quantity of drugs and with their empirically therapeutic usage'.[24] Important among these drugs, noted another French physician, was hashish, which was used as a tonic to stimulate good health rather than to fight illness and which was consumed as much as Europeans used fermented liquors.[25] In other words, as was the case in India, cannabis was seen as both a useful medicine and a pleasant intoxicant.

For as long as cannabis drugs had been used in North Africa, they had been frowned upon by the more puritanical or authoritarian of the local elites. Misgivings over intoxicants had been a feature of Islamic law and of Islamic states since the medieval period and at one point users of cannabis in Egypt had faced having their teeth pulled out as punishment for indulgence in the drug.[26] From the nineteenth century onwards the concern of the authorities had been focused on legislation. Napoleon's government in Egypt had tried to forbid by regulation the use of cannabis. The French feared that their soldiers were taking rather too freely to the local intoxicants and that this would impair their effectiveness. After the departure of the French, the Egyptian government passed its own laws in 1868[27] that were modified at regular intervals after that date. In 1879 it was decided that 'the cultivation, sale, importation or attempt to import haschisch will be punished by a fine of 200 tariff piastres per oke' and all 'boats, carriages, beasts of burden, instruments and any materials whatever which may have served for transporting haschisch, as well as any merchandise used to wrap it up in order to conceal it and facilitate its introduction shall be confiscated'.[28]

The commitment of the Egyptian government in the nineteenth century to trying to prohibit cannabis use in the local population may simply have been an updated rendering of medieval attempts to curb consumption. However, a couple of features of Egyptian history in this period may well be important in explaining the revived interest in the issue. The nineteenth century saw the rise of the government of Muhammed Ali Pasha in 1803 and from then onwards various attempts were made to reform the country and to modernise it along western lines. One of the key issues in this process of modernisation was health, as 'Muhammed Ali's drive for an effective military force, as well as for economic development, required a minimum manpower level that he soon saw threatened by serious chronic and endemic diseases'.[29] Cannabis use, which was widespread in local society, may well have been drawn into these debates about how to produce a healthy army and workforce, as indeed it was elsewhere.[30] In fact one of Ali's strategies for dealing with the health issue in the country was to found in 1827 a medical school on Western lines that was so successful that within half a century 'a medical profession had clearly come into existence in Egypt one whose members practiced modern Western medicine based on long years of studying and training'.[31] The rise of a modern medical profession in other countries was one of the key elements in increasingly hostile attitudes on the part of the authorities to unregulated consumption of drugs. In England, for example, it was the doctors that campaigned against opium substances in the nineteenth century. Preparations that were freely available for self-medication were seen as rivals to medical expertise, as individuals would not bother consulting a doctor if they felt that they could simply purchase a tried and trusted remedy over the counter that would relieve the symptoms of their suffering. As doctors formed professional bodies, these campaigned for control over such pills and potions using the argument that as well-trained practitioners they knew better than the lay person about what ought and ought not to be taken as medicine. It was for this reason that the General Medical Council and the Pharmaceutical Society saw to it that opium preparations were brought under increasing control in Victorian England.[32]

If Islamic misgivings about intoxicants, Egypt's modernising government and the rise of a westernised medical profession there provide the context of the country's laws on cannabis in the nineteenth century, they do not explain the particulars of El Guindy's spectacular performance in Geneva. After all, there had been plenty of evidence gathered that Egypt's attempts to control consumption of cannabis preparations among its population was a failure long before he presented the subject in 1924. For example, Lord Cromer wrote from Cairo in 1892 that 'the main facts of the case are very simple. On the one hand the importation and use of haschisch is forbidden by law. On the other hand, it is notorious that the drug is largely imported and very generally used. It would, under any circumstances, be very difficult to stop haschisch being smuggled into the country.'[33] At no point, however, had the Egyptian authorities tried to force it upon the agenda of the international drugs meetings as it began to emerge after 1909.

El Guindy's performance in 1924 can in fact be seen as a direct consequence of British imperial interest in the issue of international cannabis smuggling in the 1920s. After the First World War, the Government of India, the Governor of the Seychelles and the British Consul in what was then Abyssinia became caught up in one of the longest-running drug smuggling cases of the immediate post-war period. H.G. Monfreid, a Frenchman, had bought 10 tons of cannabis, probably charras, in Bombay in 1922. It was licensed for export to Abyssinia as 'an insecticide' and transported in 240 bags. The first stop on the way to Abyssinia was at the Seychelles where the British Governor immediately became suspicious. Aware that the consignment had potential as drugs, he forbade its onward journey while he investigated the subject. Having taken almost 4 months to satisfy himself that he had no grounds to further restrict its journey, he let it go on, the packages passing through French Somaliland before arriving in Abyssinia. By the time it arrived in Harar, 120 of the bags were found to contain not cannabis but organic fertiliser. The remaining 120 bags were impounded by the authorities in Abyssinia and eventually sent back to French Somaliland for collection by de Monfreid. Once he collected it there, it was shipped via Aden to Hamburg and once in Germany it was opened, the remaining 120 bags of cannabis were found to also have suddenly become organic fertiliser, 'a plant soil chemically treated on a vegetable fertilising base'. It contained large amounts of limestone and so can safely be assumed not to be a product of Indian hemp.

The mystery surrounded the question of where the cannabis had gone and how the authorities had managed to lose track of 10 tons of cannabis that they had been watching for almost 2 years between 1922 and 1924. The reason that they had been watching the consignment was that de Monfried was a notorious 'adventurer, and is reputed, probably with truth, to be engaged in gun-running and other illicit forms of enterprise'. In fact in the First World War, he was accused of supplying arms and information to the Turks in the Yemen, his wife was a German whose father still lived in Stuttgart. He had been arrested and sentenced for gunrunning in 1914/1915 and also in 1918, but on each occasion the French authorities in Somaliland vouched for him and secured his freedom.

De Monfreid argued that he was in fact the victim of deception.[34] Having exported the Indian hemp as an insecticide commonly used in Turkestan, a gang of Greeks was supposed to have stolen some and smuggled it into Egypt on board the ship of a passing princess, while de Monfried's agent Captain Ternel was alleged to have tried to sell the consignment to an Egyptian syndicate after attempting to bribe the Italian and the French consuls on the Seychelles for shipping permits. The British, however, had other evidence. In 1925 the Consulate in Abyssinia wrote that the Frenchman had made such significant profits from cannabis running back in 1923 that he had been able to buy a controlling interest in the Flour Mills and Electric Lighting Works in Dire Dawa. Indeed, Monfreid had begun to encourage local farmers in Abyssinia to grow Indian hemp for him. Based in French Somaliland, he collected the locally grown cannabis and added it to that which he was obtaining from elsewhere. He then had this raw hemp manufactured into charas and then arranged to smuggle it into Egypt. The British Consul writing in 1925 observed that 'de Monfreid's initial essays in this respect two years ago were so successful that it is hard to believe that he would easily abandon such a lucrative adventure'.

Interest in this case was such that the Foreign Office compiled a dossier on it and regularly supplied information to the British administrators in Egypt itself. Indeed, because of this stream of information from across the empire about cannabis smugglers it may well have been them who alerted the local authorities there to the activities of de Monfreid, as the Egyptian police reported in March 1924 that acting 'on information supplied to them', they had succeeded in thwarting some but not all of the smuggling activities. Aware that Egypt was caught up in this large-scale illicit trade, the British in Cairo decided to make their own assessment of the situation there. In February 1924 the European Department of the Ministry of the Interior stated that 'the desirability of international cooperation for the control of hashish led to investigations being made in Egypt on the whole question of hashish traffic'. A report was compiled that Lord Allenby forwarded to London in April 1924.

The document was a lengthy investigation of the extent of cannabis use in Egypt. It suggested the complete failure of existing regulations to have an impact on hashish consumption in the country. The key problem identified was the ease of supply. The routes by which the drug came to be in the country were varied and difficult to detect. The report noted, for example, that 'when sent on steamers from Greece and Syria it is put in rubber bags and dropped overboard near the coast: the bags are marked by a float and are picked up by agents afterwards' and it had also found that 'along the coast west of Alexandria or East of Port Said it is landed in small boats and carried inland by Bedouin on camels'. There was a hint of despair as the report noted the many ways in which the drug found its way into the country, 'hashish has been found in the middle of cotton goods from Manchester, it is put in bundles of newspapers from abroad, it is hidden inside imported goods of all kinds such as tins of petroleum, bricks, millstones, marble columns, hollow bedsteads, barrels of olives, looking glasses etc.' and even 'sometimes it is made up to resemble the inner soles of boots and the smugglers merely walk ashore with it'. The drug originated in India, Turkey, Greece and Syria and once in Egypt it was consumed in a variety of ways. The ordinary folk smoked it mixed with molasses and tobacco in a 'goza' or pipe made of coconut-shell. The better off smoked it in cigarettes or ate it as a sweet mixed with spices and honey or sugar. Occasionally it was taken in coffee.

A number of agencies were responsible for policing the trade in the drugs, including the Coast Guard, the Customs officials and the City Police. The failure of these groups to prevent the trade was blamed on the size of Egypt's borders, the ease with which the drugs were concealed and on corruption; 'that collusion exists between the smugglers and some of the employees of the above administrations can hardly be denied; nor is it surprising in view of the enormous profits accruing from the sale of hashish'. Indeed, even where smugglers were caught the penalties seemed insufficient, and the report felt that there were special problems in cases where the criminals were foreigners. Under colonial regulations, the Western suspect was subject not to Egyptian courts but to the jurisdiction of his country's consular court. The report diplomatically noted that in these circumstances 'it is believed that a condemnation is difficult to obtain'.

The conclusion of the report made it clear that 'it is practically impossible to keep hashish out of Egypt'. It was adamant that the only way to improve the situation was to focus on stopping the supply from getting to the country in the first place. The onus was put on international cooperation to achieve this. In the first place, the report advocated approaching the Capitulary Powers in order to have their subjects subjected to the same laws as for Egyptians. Secondly it noted that 'the import of hashish into Egypt might be reduced if the question was taken up by the League of Nations and hashish was included in the International Opium Convention [as] it is extremely difficult for one country like Egypt to fight single-handed against a noxious drug the traffic which is openly tolerated in other countries'.[35] The report ended on the conclusion that 'it is suggested therefore that the League of Nations should consider hashish traffic as an international affair and should try to persuade its members to make dealing in or consuming the drug a crime punishable by severe penalties'.

It is worth bearing in mind the fact that these recommendations almost exactly anticipated El Guindy's agenda, and even his language, at the Second Conference that itself came just over six months after the report. It is also important to consider the fact that the report was made available to Egyptian administrators as it was forwarded to the Director General of Public Security in Cairo. In this light it seems likely that this is the source of the Egyptian assault on the Geneva conference in 1924. Indeed, one of the reasons why El Guindy and his delegation would have seized on the report with such enthusiasm is provided in a cautionary note scribbled on the Foreign Office copy of the report in London. Malcolm Delevingne, the British member of the Advisory Committee, saw this despatch from Lord Allenby and noted that it was 'a very interesting report'. However, he scribbled the argument on the cover of the dossier that it was 'one that for the present we can keep to ourselves'. His reason for this was that he had no intention of providing 'a stick to beat the Capitulary Powers' to those opposed to European interference in Egyptian affairs. The British, who had after all governed Egypt directly between 1882 and 1922 and who retained an informal presence in the country's affairs until after the Second World War, were perfectly aware of the fact that the country was being swamped with smuggled cannabis and that its laws on the subject throughout this period had failed, and yet they had not bothered to do much about it. In a political climate where Britain's colonial record on drugs was the subject of such scrutiny, another failure would damage both the moral authority of the imperial presence in Egypt and the bargaining position of the British in the international drugs debates. El Guindy's focus on the cannabis issue at the Second Opium Conference may well have stemmed from a genuine concern about the drug. However, many of his facts and most of his agenda seem to have come from a British report and his enthusiasm for the subject would have been bolstered by the knowledge that the issue was one that would damage his country's former colonial masters. He had, after all, made a point of announcing his presence at the Opium Conference with the pointed reminder that 'this is the first time that Egypt had been represented by a purely Egyptian delegation at an international conference'.[36]

The argument that the position taken by El Guindy at Geneva was actually founded on British sources is made even more compelling when the scientific basis of the Egyptian argument is considered. El Guindy's dramatic announcements on the mental health implications of cannabis use in Egypt had had a considerable impact on the assembled delegates as he was able to support these with statistics, 'illicit use of hashish is the principal cause of most of the cases of insanity occurring in Egypt... generally speaking, the proportion of cases of insanity caused by the use of hashish varies from 30 to 60 per cent of the total number occurring in Egypt'. Similar evidence made up part of the official 'Memorandum with reference to Haschiche as it concerns Egypt' that was submitted by the delegation in support of El Guindy's speeches. However, this time the figure was even more alarming 'about 70% insane people in lunatic asylums in Egypt are haschiche eaters or smokers'.[37]

In all of the Egyptian campaigning this was the only material that was produced that might be considered 'medical' or 'scientific' evidence. It is interesting to note then that the Egyptian Lunacy Department, from which these statistics would have been drawn, had been the personal fiefdom of an Englishman for over a quarter of a century between 1895 and 1923. John Warnock was appointed by the Public Health Department in Cairo in 1895 at a time when Egypt was an established part of the British empire. He had been working in the British asylum system for almost a decade by this time and was seen as the ideal man to reform the Abbasiya Asylum. He remained at the task for 28 years. During this time, he expanded the existing institution, built a new hospital, drafted laws on mental illness in Egypt, and created a whole new Department dedicated to Lunacy within the colonial Ministry of the Interior. By the time he retired almost 2500 Egyptians were being treated at any one time within the units of the Lunacy Department.

He seems to have developed little attachment to the place that was to be his home for such a large part of his life. He admitted that he did not study written Arabic and that he found it 'impossible to learn all the tongues necessary to converse with all the patients and their friends' and his grasp of the vernacular was such that he could only 'make my wants known and give orders'. The country exhausted him, and by 1916 he had to take a long leave from the stress of work in Egypt that had been exacerbated by the presence of shell-shocked soldiers from the African campaigns of the First World War. He contemptuously dimissed Egyptian political ambitions after the First World War and noted that 'self-determination was proving to be an infectious mental disorder'.[38] Yet despite this apparent lack of sympathy with the society around him, he felt sure that he could locate the chief cause of insanity in the Egyptian population. This was the use of cannabis.

His first year at the asylum was a particularly trying period. He arrived in February 1895 and noted the following difficulties.

> Besides the almost complete lack of funds, my total ignorance of Arabic, and the total ignorance of patients and staff of any language but Arabic, prevented my doing anything for some time. I was unable even to tell the servant to shut the door or to ask a patient his name. I had no interpreter. However, after some time I found a patient who could write English and for a while he was employed in translating Arabic letters etc until it was discovered that he interpolated numerous mis-statements founded on his delusions. In those days an English or French-speaking clerk was not available. For a time I could only look on and guess at what was going on in most matters.

Yet despite the range of difficulties in gathering accurate details about patients that included problems of translation, problems of deliberate mis-information, problems of communicating with staff and a reliance on guess work, Warnock claimed that he was able to produce an authoritative account of the causes of mental illness in the asylum within 10 months of his arrival. This was reported as 'The Cairo Asylum: Dr Warnock on Hasheesh Insanity by TS Clouston MD Edinburgh' that was published in the *Journal of Mental Science* in 1896. This was a summary of Warnock's observations of the asylum statistics that his hospital had generated in the period from his arrival at the hospital in

February to the end of 1895. These statistics were central to the argument and after noting such numbers as 'in 41 percent of all his male patients hasheesh alone or combined with alcohol caused the disease', he concluded that 'I have no doubt that in quite a number of cases there hasheesh is the chief if not the only cause of the mental disease'. He went on to note the clinical features of this 'Hasheesh Insanity' that included 'an elated, reckless state, in which optical hallucinations and delusions that devils possess the subject frequently exist' or even 'terrifying hallucinations, fear of neighbours, outrageous conduct, continual restlessness and talking, sleeplessness, exhaustion, marked incoherence and complete absorption in insane ideas'. The statistics, and the exotic location, seem to have been enough to convince TS Clouston who exclaimed 'such are the latest words in regard to hasheesh and its insanity'.[39]

Despite Warnock's frank admissions that he had very little idea of what was going on upon his arrival in Egypt and indeed had no reliable means of remedying this situation beyond hazarding a few guesses of his own and trying to interpret the lunatic translations of his delusional clerk, it seems that he was happy to jump to conclusions about the cause of illness among a large proportion of his patients within 12 months of his taking up the post. He may well have read an earlier report on Egyptian mental illness, he had certainly seen this by the end of his career as he noted it in his 1924 article, which argued that 'with the men the attack of insanity was attributed in nearly all cases to one of three causes, the use of hashish, some disappointment or grief, and religious excitement. Of these, the first is by far the most frequent.'[40] Whatever was the case, these were conclusions that he stuck to. In 1903, he published a lengthy account of his observations at the asylum. Again he relied on numerical evidence to make his point, 'in Egypt, statistics are available since the year 1895. During the six years 1896–1901 out of 2564 male cases of insanity admitted to the Egyptian Asylum at Cairo, 689 were attributed to the abuse of hasheesh, i.e. nearly 27 per cent.' He quoted statistics from India to make the comparison, 'between 1882 and 1892 Indian hemp caused 25 to 35 per cent of the insanity in Bengal asylums' even though the reliability of these numbers had been challenged by the Indian Hemp Drugs Commission itself (IHDC). He was at pains to refute the conclusions of the IHDC and emphasised that 'my experience does not confirm the Indian Commission's belief that cannabis indica only sometimes causes insanity. In Egypt it frequently causes insanity'. He was keen to stress that his statistics, and remember that they were his statistics as he admits that the collection of this data only began in 1895 when he arrived, were entirely dependable. He did this by claiming that each patient counted as a sufferer of hasheesh insanity was correctly diagnosed. He did not believe police reports of hasheesh use nor did he give much credence to relatives of the patient. Indeed, he did not believe the patients themselves noting that 'excited protests and denials of the habit are known by experience to indicate a hardened hasheesh smoker'. Instead he relied on his own intuition and repeated questioning of the patients until a confession was obtained.

Quite how reliable this method was of establishing that a case was one of cannabis use is worth considering. In 1895, he stated that he thought that one of the key symptoms of weak mindedness caused by hasheesh insanity was that 'they deny the use of hasheesh'. He made it clear in 1903 that 'as the mental state of the patient improves he is again questioned about hasheesh and before discharge he is invited to give full details of his habit'. It seems then that procedures in Warnock's hospital encouraged inmates to confess the use of cannabis preparations, as the final hurdle before release was another interrogation on the subject of cannabis use by a doctor who admitted that he could consider a denial of the habit as a symptom of problems of mental illness.

In fact his conclusions themselves are more wide ranging than they have a right to be. Based on his experience of cases at the asylum that he believed to be caused by cannabis use, he made sweeping observations such as 'the use of cannabis indica in Egypt seems to have graver mental and social results than in India and is responsible for a large amount of insanity and crime in this country'. However, he also admitted that 'as to whether excessive use of hemp drugs is commoner here than in India I can give no opinion, but many thousands use it daily here' and indeed went further in noting that while 'many thousands smoke hasheesh only a comparatively few suffer from grave toxic symptoms'.[41] In other words, he made broad generalisations about cannabis use and cannabis users that were meant to apply to all users in all of Egypt despite the fact that he saw only a small proportion of them at the hospitals. The issue of whether this was a representative proportion of the cannabis users in the country never seems to have troubled him and he broadened his conclusions drawn from the troubled individuals at the asylum to apply to thousands of ordinary Egyptians that took hasheesh and yet never became subjects of his scrutiny. In short, his method of establishing that an individual at his hospital was a cannabis user was suspect and the conclusions that he drew about cannabis use in general were based simply on the small sample of all of Egypt's many users that had ended up in his hospital. Much as in India in the nineteenth century, the habits indulged in by much of the local population were condemned by colonial doctors who had no idea what was going on outside of the walls of the hospital and to whom it never occurred that a small band of lunatics could in no way be considered a representative sample on which to base observations about wider society.

Conclusion

It is important to recall that these were the circumstances in which cannabis first found its way into the concerns of the international bodies and systems that were emerging in the 1920s to govern intoxicating drugs. The matter was first raised at The Hague by South Africa, a country in which anxieties about cannabis had been growing since the 1880s. These anxieties show that cannabis had become intertwined with the complex forces that shaped Natal, as the colony's white elites associated it with reluctant workers and those who transcended the economic and racial barriers that they found it expedient to erect. The issue was subsequently forced on the conference by Egyptian delegates eager to dent Britain's international standing because it was their former colonial ruler. However, they used the language and arguments of a British report (that should have been covered up) and they were armed with questionable statistics from a mental health system organised by those foreign imperialists and under the charge of a colonial doctor that had little knowledge of Egyptian society.

The irony of all this is that cannabis was forced on to the agenda at The Hague despite the opposition of the British representatives there. They had tried to keep cannabis off of the agenda, and had found themselves forced to exempt India from controls on the drug as they derived substantial revenues from taxing consumption there. Ultimately their opposition and obstruction were futile, and cannabis found itself scheduled included in the nascent system of controls on intoxicating drugs, because of the complex politics of British imperialism at either end of Africa.

Notes

1. W. Walker, *Opium and Foreign Policy: The Anglo-American Search for Order in Asia, 1912–1954* (Chapel Hill: University of North Carolina Press 1991); W. McAllister, *Drug Diplomacy in the Twentieth Century* (London: Routledge 2000); A. Taylor, *American Diplomacy and the Narcotics Traffic 1900–1939* (Durham: Duke University Press 1969); Timothy Brook and Bob Tadashi Wakabayashi (eds), *Opium Regimes: China, Britain and Japan, 1839–1952*

(Berkeley: University of California Press 2000); C. Trocki, *Opium, Empire and the Global Political Economy* (London: Routledge 1999).
2. Report of the British Delegates to the International Opium Conference held at the Hague, December 1911–January 1912 (London: HMSO 1912), p. 14.
3. For more on this episode, see McAllister, *Drug Diplomacy*, p. 28; Taylor, *American* Diplomacy, p. 87; J. Mills, *Cannabis Britannica: Empire, Trade and Prohibition* (Oxford: Oxford University Press 2003), p. 88.
4. The letter confused 'dagga' for 'dakka' in the language of local drug users in South Africa. The former was the African word for the leaves of the Leonotus plants that were smoked, the latter the local slang for cannabis preparations.
5. Minutes of the sixth session of the Advisory Committee on traffic in opium and other dangerous drugs 12.08.1924, p. 52. This case will be considered in more detail later in the chapter.
6. Records of the Second Opium Conference v. 1 20.11.1924, pp. 39–40.
7. For a full account of how the Egyptians steered cannabis through this conference and on to its treaties, see Mills, *Cannabis Britanica*, pp. 152–87.
8. S. Bhana, *Indentured Indian Emigrants to Natal, 1860–1902* (New Delhi: Promilla 1991), pp. 7–17.
9. Report of the Indian Immigrants Commission (Natal Legislative Council) 1885–1887 (Pietermaritzburg: Davis 1887), p. 7. Hereafter IIC.
10. Ibid., p. 244; As stated above, dakkha was the local African word for cannabis.
11. Ibid., p. 405.
12. Ibid., p. 410.
13. Ibid., p. 7.
14. These numbers are taken from India Office Library L/PJ/6/284/1476 Report on the IIC.
15. IIC, p. 282.
16. Ibid. p. 74.
17. Ibid., pp. 43 and 175.
18. Ibid., p. 6.
19. B.M. du Toit, 'Dagga: The History and Ethnographic Setting of Cannabis Sativa in Southern Africa' in V. Rubin (ed.), *Cannabis and Culture* (Paris: Mouton Publishers 1975), pp. 81–116.
20. IIC, p. 209.
21. Ibid., p. 408.
22. Ibid., p. 410.
23. Ibid., p. 8.
24. Charles Cuny writing in 1853 quoted in L. Kuhnke, *Lives at Risk: Public Health in Nineteenth Century Egypt* (Berkeley: University of California Press 1990), p. 28.
25. Pierre-Charles Rouyer quoted in Ibid., p. 29.
26. A.M. Khalifa, 'Traditional patterns of hashish use in Egypt' in Rubin, *Cannabis and Culture*, p. 199. The problem with an understanding of exactly what Islamic legislators and jurists were discussing when it comes to narcotics and stimulants is that the words 'Hashish' (cannabis sativa), 'banj' (drugs) and 'afyūn' (opium) are often lumped together without a precise definition of these terms. Hashish is not, however, mentioned in the Qur'an or indeed in the oldest Islamic legal texts. For details of this, see F. Rosenthal, *The Herb: Hashish Versus Medieval Muslim Society* (Leiden: Brill 1971).

27. 'The Policy of Indian Hemp Drug Administration' in Report of the Indian Hemp Drugs Commission, volume 1 (Simla 1894), p. 270.
28. 'Government of Egypt Decree of March 10 1884' in Reports from Her Majesty's Representatives in Egypt, Greece and Turkey on Regulations affecting the Importation and Sale of Haschisch, in Parliamentary Papers LXXXIX, 1893–4, p. 301.
29. Kuhnke, *Lives at Risk*, p. 12.
30. See the discussion of South Africa below.
31. A. El Azhary Sonbol, *The Creation of a Medical Profession in Egypt, 1800–1922* (New York: Syracuse University Press 1991), p. 134
32. See V. Berridge, *Opium and the People: Opiate Use and Drug Control Policy in Nineteenth and Early Twentieth Century England* (London: Free Association Books 1999), pp. 113–22.
33. Reports from Her Majesty's Representatives, p. 294.
34. In fact de Monfreid later accused the British of hounding him in order to protect its clandestine interest in monopolising smuggling into Egypt. See H. de Monfreid (translated by Helen Buchanan Bell), *Hashish : A Smuggler's Tale* (London: Pimlico 1994).
35. NA HO 144/6073 Hashish: Traffic and consumption in Egypt 02.04.1924.
36. Records of the Second Opium Conference v. 1 20.11.1924, pp. 39–40.
37. NA HO 144/6073 Egyptian Proposal for Inclusion of Hashish 12.12.1924.
38. J. Warnock, 'Twenty-eight years' Lunacy Experience in Egypt (1895–1923)' in *Journal of Mental Science*, lxx, 1924, pp. 233–61.
39. T. Clouston, 'The Cairo Asylum: Dr Warnock on Hasheesh Insanity' in *Journal of Mental Science*, 42, 1896, pp. 793–4.
40. A. Urquhart and S. Tuke, 'Two visits to the Cairo Asylum, 1877 and 1878' in *Journal of Mental Science*, 25, 1879–80, pp. 43–53.
41. J. Warnock, 'Insanity from Hasheesh', in *Journal of Mental Science*, 49, 1903, pp. 96–110.
42. For more on British positions at The Hague, see Mills, *Cannabis Britannica*, pp. 165–77.

11
'A Grave Danger to the Peace of the East': Opium and Imperial Rivalry in China, 1895–1920

William O. Walker III

Introduction

The British presence in China focused largely upon commercial and financial interests that were increasingly threatened by the turn of the nineteenth century. Maintaining India was a costly affair and as Britain watched other, larger industrial nations like Germany develop an enormous productive capacity in the late nineteenth century, British officials felt that only domination of global finance and a towering presence in international shipping would sustain their nation's great-power status. Thus, active and successful involvement in China was a virtual *sine qua non* of British identity. Japan had both less and more at stake in China; Japan could not hope to match the British financially or commercially in Asia at this time, but knew that a failure to establish a prominent foothold in northern China and Manchuria would fatally undermine Japan's plans to attain great-power status through the development of regional strength.

The Spanish–American War gave the United States its own colony in Asia, the Philippine Islands, and with the announcement of the Open Door policy in the forceful diplomatic notes of Secretary of State John Hay in 1899 and 1900, the third interest in the region was declared. As American interests in China came into focus, a grandiose ambition was revealed. Acting in concert, American diplomats, merchants, and missionaries would stop at nothing short of the cultural transformation of China into a land receptive to the best the United States had to offer. This included American-made industrial and agricultural products, American-style Christianity and a liberal form of government. Realization of this goal, which one scholar has termed 'beneficent imperialism' would certainly pose a threat to competing imperial interests in China.[1]

This chapter examines the ways in which opium consumption in China, and the trading networks that supplied the market there, became caught up in the politics of imperial rivalry between three powers in East Asia in the first decades of the twentieth century. It was the competing ambitions of Britain, Japan, and the United States in the region that shaped the ways in which the opium issue was formulated, approached, and handled.

To be sure, empire and the rivalries that arise therefrom begin in large part as outcomes of an organic process of self-perception. Identity for great powers begets empire.[2] In the concrete shapes that imperial identity ultimately assumes, there are mediating factors. On the imperial stage that was China and in the rivalries that played out on that stage, one such factor was opium and the opiates, like morphine and heroin, that are the by-product of the opium poppy. No recounting of the imperial contest that occurred in East Asia early in the 1900s would be complete without an understanding of the place of opium in that contest. During that time, opium proved to be something of an obstacle to the realization of imperial aspirations.

An inquiry into the role of opium in China between 1895 and 1920 necessitates the adoption of a viable analytical framework. The narrative of this story itself is, at the least, a four-sided, nationally based construction that incorporates the interests of Great Britain, the United States, Japan, and China. Bringing conceptual order to this situation entails looking at the period from three perspectives: reform, security, and governance. The fundamental issues crystallized around the questions of how best to reduce opium's effect upon China's populace and to reduce reliance upon revenue derived from its sale; how best to provide China with a secure buffer against assaults by opium traders; and how best to fashion a model of governance for China that would not have to focus inordinately upon the issue of opium control.

The turn of the century context

The need for opium reform seemed undeniable by the late nineteenth century. Trade in opium from India and the Ottoman Empire had combined by about 1830 to make rampant corruption appear to be the natural order of administration in Chinese ports. Subsequent proclamations against the trade and against opium smoking became little more than ruses to assure badly paid officials a reliable income. On the eve of the first Opium War, there were as many as 10 million inveterate users of opium.[3] With the increasing scarcity of silver, taxes on opium gave the dynasty an indispensable source of revenue. Commissioner Lin Zexu's failure to put an end to the trade, as the Daoguang Emperor had instructed him, showed how endemic the opium problem actually was in China. Great Britain, hewing to its nascent foreign policy of free trade, did not forbid its vessels from carrying a nominally illicit commodity, opium, from India into Chinese waters. There was no motivation to do so as the Manchus rarely strove to enforce existing antiopium legislation. Moreover, as popular demand for opium increased, domestic production began to flourish by mid-century in Yunnan, Guizhou, and Sichuan.[4]

With the cessation of the second Opium War in 1860, the British were free to sell opium at the treaty ports, where it was happily received as a taxable commodity. This change occurred as dynastic officials began to cede authority, however unwillingly, to regional elites, landlords, and militarists. This ostensible legalization of the opium trade coincided as well with the decision of Great Britain in the early 1860s to participate in the suppression of the Taiping Rebellion, no charitable gesture given Taiping strictures against the use of opium. Curtailing the trade in the name of reform would have been detrimental to British colonial policy and, hence, could have affected Britain's international status by threatening a symbolically and literally important source of the Crown's revenue.[5]

The outcome of the Sino-Japanese War in 1895 set the stage for the struggle against opium in China in the early twentieth century. The weakened Qing dynasty ceded Formosa to Japan and formally recognized the independence of Korea, which not only deprived China of revenue but also further impaired dynastic authority both at home and in the eyes of foreign powers. In the process, the incidence of opium use grew at such an alarming rate that by the turn of the century perhaps one-fourth of all Chinese smoked prepared opium.[6]

Nearly 25 years earlier, China's first envoy to London, Guo Songdao, complained that Westerners did not care about opium's impact upon China. 'For several decades', he noted, 'it has been the national humiliation, it has exhausted our financial power and poisoned and injured the lives of our people, but there is not a single person whose conscience is weighed down by it.'[7] Here was the argument that the roots of China's troubles lay abroad, a theme common to the entreaties of drug prohibitionists throughout the twentieth century. It was not only greedy Westerners, though, who were responsible for China's misfortunes with opium. No one of any nationality who profited from the trade in opium would willingly consent to preside over its demise. Foreign missionaries blamed the Chinese themselves for the many troubles associated with opium in light of the dynasty's apparent tolerance of poppy cultivation. By 1895 the United States, alone among the powers, forbade its nationals from participating in the opium business.[8]

The report of the British Royal Opium Commission in that same year held beleaguered Qing officials partly responsible for failing to curb the lucrative trade. Nevertheless, a dissenting minority argued that it was wrong for 'the British Empire to be engaged in a traffic which produces such widespread misery and disaster'.[9] The internal crisis of the late 1890s, which helped to spark the Boxer uprising, portended the possible break-up of China before giving strength to an incipient reform movement. At the forefront of that movement was the reformer Zhang Zhidong who was struggling to find a Chinese road to reform. Opium control was one means he called upon to restore the integrity of the dynasty. 'The development of education', contended Zhang, 'is the medicine to use for opium suppression.'[10] A stricter and punitive tax policy on raw and prepared opium, which was presumably to be applied to all of China's provinces in 1906, was to accompany other means of reform.[11]

Despite subsequent claims to the contrary by Western missionaries, which were made as if to remind the Qing of the identity of their benefactors, the opium reform movement was therefore initially of Chinese origin. Chinese students returning from Japan began calling for reforms after watching a government monopoly there and on Formosa reduce the incidence of opium smoking. In 1906 the dynasty bowed to this and similar sentiments by issuing an edict that mandated the cessation of opium cultivation over a 10-year period and required, as did the Formosan model, licenses for smokers. One noted observer of China, H. B. Morse, commented that the 'effect on the nation was electrical'.[12]

Asia's opium politics in the early twentieth century

The next step in reform was to put an end to the commerce in Indian opium, something favored by antiopium societies in both Britain and China. Despite misgivings about the sincerity of each other's professed commitment, the two nations agreed on a trial basis in late 1907 to reduce the trade in prepared opium by 10 per cent annually. The base figure was derived from the average annual volume of opium exported to China between 1901 and 1905. In a decidedly imperial twist, the monitors of the agreement in the main would be foreign missionaries and British consular officials.[13]

This crackdown on opium came at a time when the Qing dynasty was especially vulnerable to internal rupture. Consequently, a companion-mandated decrease in the domestic production of opium meant severe economic hardship for those provinces that had previously relied upon opium as a revenue-producing crop. Those hit the hardest included Shaanxi, Guizhou, Yunnan, Sichuan, and to a lesser extent Manchuria where opium had been growing for 50 years. Perhaps surprisingly for a country on the verge of revolution, China compiled a strong enough record in 3 years, so that in 1911 the Sino-British agreement was extended through 1917, the initial target date. Appeals from antiopium associations in Britain and from the Chinese National Anti-Opium Society to end the trade sooner did not succeed.[14]

Meanwhile the United States, which had undertaken a campaign a few years earlier to minimize the incidence of opium smoking in its own colonial possession, the Philippine Islands, pressed for a more comprehensive antiopium accord that would cover all of East and Southeast Asia. The agreement to end the opium trade in Asia was to be reached at a special conference. However, officials in Washington who proposed the meeting did not seriously consider what that possible development could mean for Chinese integrity or, perhaps more importantly, for how other powers with economic and strategic interests in East Asia would assess the relative strength of the Qing. The prospect of opium reform in the American style trumped all such considerations.

The US effort to set an agenda of proscriptive reform for the Shanghai Opium Commission in 1909 failed. It is worth noting that American planners did not ignore Japan as they prepared for the antiopium conference. Yet to their chagrin, the Japanese were not yet commonly seen by Western powers as fully equal actors in all affairs pertaining to international relations in East Asia.[15] That misperception was, of course, about to change. Shanghai provided a lens on the near future because morphine exports from Japan to China, though low in volume, had caught the attention of US and other consular officials. While at the inception of the conference, in February, Japan had not enacted prohibitions on the sale of morphine or needles for injection, by the end of the year such a prohibition was in place: a development that indicated just how eager the Japanese were at that time to protect their image, particularly as their regional aspirations were on the rise.[16]

Since the conference was only a commission and not a formal parley whose participants possessed full diplomatic powers, the delegates could only make non-binding recommendations about how to address the opium problem in Asia. This lack of authority derived in large measure from a British reluctance fully to abandon revenue from the opium trade and also from Whitehall's refusal to reconsider the terms of the unequal treaties, which had been forced upon China in the aftermath of the Opium Wars. The meeting at Shanghai therefore accomplished little of substance. Resolutions expressed gratitude for China's antiopium efforts, called for the convening of additional such gatherings, and stressed the need for producer states to curb the export of opiates.[17]

One result of the conference is pertinent for present purposes: Shanghai brought out Anglo-American differences, a putative imperial rivalry, over how best to control opium. Embedded within those differences were questions about to which nation, Britain or America, China would turn as it continued its path toward becoming a modern state. The British clung to their belief, however implicit it may have been, that China was not a full-fledged state since it could not halt the cultivation of or traffic in opium. Indeed, since commerce in opium was still permissible, the British feared that domestically produced opium would hurt the market for Indian opium and as such would hurt the financial arrangements of maintaining their empire. London would only be willing to countenance such a drastic change in colonial administration if China demonstrated that it could be trusted to carry out full prohibition on poppy growth and the illicit commerce in narcotics.[18]

The United States sensed that expressions of concern for China's welfare would pay dividends on a regional level; this sentiment was especially in vogue during the presidency of William H. Taft.[19] Washington therefore sought to move the discussion of opium control to the international arena where Great Britain's influence might be better contained. As such the United States, with the support of the Qing, asked interested governments to attend an antiopium conference that would convene at The Hague in late 1911. British Foreign Secretary Sir Edward Grey delayed accepting the invitation for reasons just adduced; a similar reluctance emanated from Tokyo which wanted a guarantee that local morphine manufacturers would be able to compete in the manifestly lucrative market for opiates.[20]

Less than 2 months before the start of the conference, China was thrown into turmoil with the start of the 1911 Revolution. The decentralization that had affected governance in China since the Taiping Rebellion explains the broad movement that joined provincial gentry, merchants, and militarists against the dynasty. On February 13, 1912, Sun Yat-sen turned over to Yuan Shihkai the reins of the new Chinese Republic.[21] No longer were the deliberations at The Hague primarily concerned with the nature and pace of opium reform in a China that was disintegrating; they were now directly related to the issue of the integrity of the new country itself. The authorities of the new China displayed their own imperial ambitions at The Hague by signing an accord not only for China proper but also in the name of China's territories, which they assumed included Tibet despite a proclamation of independence by the Dalai Lama in Lhasa.[22]

Thus it was that security and opium reform became indivisible even though the exact nature and timing of reform remained hotly contested. The United States tried and failed to set binding principles for controlling the trade in raw and prepared opium and also failed to get agreement on the need for a special administrative body to oversee the progress of opium reform. The British report on the conference at The Hague denounced the US delegation for introducing 'a number of useless and irritating proposals'. As for the Chinese, they 'constantly endeavour[ed] to pose as the protagonists in the matter of opium reform and to divert attention from their own shortcomings'. The evident lack of substance in the British report on the conference suggests an awareness of the attack on Whitehall's policy of gradualism on opium reform. The British, too, were noticeably upset because the Japanese seemed to have their interests in the morphine trade protected by provisions in the resultant 1912 Hague Convention.[23]

Japan's diplomatic success in a limited role at The Hague reflected the coming change in its stature as a regional power. Alliance agreements with Great Britain signed in 1902, 1905 and 1911 recognized Japan's strategic interests and commanding position in East Asia, while assuming a comity of political objectives with the British.[24] Having been perceived by the British as their junior partners, and thus denied equal status in East Asia, the 1912 Hague Convention suggested otherwise. It was becoming abundantly clear that Japan, however selective its role at The Hague may have been, was prepared avidly to pursue and protect its own economic interests, interests that included profits from the commerce in morphine.[25] It was also the case that Tokyo's economic goals were virtually inseparable from its larger foreign policy objectives. What this might ultimately mean for Chinese security could not be known at the close of the gathering at The Hague. What was clear, however, was that the trade in opium, the commerce in opiates, and the fate of opium reform were now, to a considerable extent, crucial arbiters of China's security. Tokyo would surely influence how those matters would play out in East Asia.

Moreover, the issues of security and governance, though not exactly parallel, were nevertheless closely linked. Governing the young state would be difficult for a number of reasons. There were many competing ideas of what post-Qing China should look like administratively as well as politically and economically. The role of the many provinces and the militarists therein remained unclear. No one could know what the aftermath of revolution would mean as the foreign powers sought to protect their economic interests and extraterritorial rights. On the eve of the Great War, which soon would pose a substantial threat to the integrity of China if only because it emboldened the Japanese to consolidate their vital interests there, China's fate did not primarily rest in the hands of the Chinese themselves. The potential of narcotics to further complicate matters of security and governance for the young republic could not be ignored.

Japan's concerns as the revolution unfolded did not augur well for China's future. The authorities in Tokyo had initially hoped for the creation of a constitutional monarchy over which it might exert preeminent influence among the interested powers. The very idea of an independent republic, however, seemed to threaten Japan's strategic, financial, and commercial interests. Yuan Shikai briefly allayed Japan's concerns and those of other powers when he made it clear that his government would bring no sudden or dramatic change to China's place in the East Asian order.[26]

Where, it must be asked, did opium fit as Japan re-thought its China policy? The Formosan experience convinced the Japanese that the Chinese had an affinity for opiates; opium smokers on Formosa came primarily from the province of Fujian and the Hakka minority in Guangdong.[27] It was widely believed that opium's hold upon Chinese society would not soon be broken. Hence, by continuing to sell derivatives of opium in China, Japanese merchants were simply responding to market conditions, as had the British before them. That there might be political and strategic repercussions to economic policy was assumed if not always discussed. Also, the very sale of opiates manifested Japan's sense of cultural superiority, as the presence of foreign concessions throughout China's ports and major cities also evinced. China's hope for acceptance as a modern nation, therefore, rested in part upon the Republic's response to opium. Yet so multi-layered was the opium issue that success in prohibition or restriction threatened Japanese ambitions in East Asia and also ran counter to British economic interests and, to an extent, Whitehall's larger imperial designs.

Yuan Shikai came to the presidency of the Republic with a credible record as an antiopium reformer. As governor-general of Zhili in the early 1900s, he controlled public affairs so effectively that the region was regarded as something of a model province in the late Qing era. He managed to cut the production of opium and establish, with the help of merchants in Tianjin, a provincial monopoly for its sale. The proceeds of opium and narcotics sales were distributed to the government and merchants on a proportional basis.[28] As president, he endeavored to continue this rationalist approach to opium reform. In late October 1912, Yuan declared his opposition to 'the all-destroying weed... which threatens to annihilate both our country and our race'.[29]

Pious proclamations, though, were no substitute for effective opium control. The need to achieve significant opium reform was never clearer than when the British all but made recognition of the Republic dependent on opium suppression: such was the position of British Foreign Minister Sir Edward Grey as conveyed to Minister Sir John Jordan in Beijing. Jordan, whose knowledge of China was substantial and whose respect for Yuan Shikai was genuine, informed Whitehall that 'there is no reason to doubt the sincerity of the President'. Consequently, Great Britain extended recognition to the Republic in a timely fashion.[30]

Opium poppy cultivation, however, remained a serious problem in China. In Yunnan, landlords ordered peasants to grow opium and the new government exercised little authority in regions of Sichuan where cultivation flourished; the situation was similar in Gansu. Jordan nevertheless reported to his superiors that 'measures are being adopted nearly everywhere to suppress the cultivation, trade, and consumption of opium, and there is no evidence of any discrimination in favour of the native drug'. In line with Britain's own preference for suppression, Jordan applauded the government's plan gradually to curtail financial reliance on opium.[31]

The British were not prepared, however, simply to walk away from the economic benefits attendant to the opium trade in China. Revenue from whatever source inevitably contributed to the health of the empire. So it was that the British Government became a proponent of a consortium loan for the financial reorganization of China in the aftermath of the 1911 revolution.[32] In co-sponsoring the project, the British sought to perpetuate China's financial, if not diplomatic, indebtedness to Great Britain. In Whitehall's thinking, indebtedness meant the strengthening of British-Sino commercial relations, which would have the added advantage of checking the rapid growth of a Japanese commercial presence in north China.

For its part, Japan could not have welcomed Yuan's campaign to suppress opium with the same enthusiasm as that of Great Britain. The more successful China's antiopium campaign became, the more imperiled would be Japan's imperial intentions. A stable China, devoid of its traditional addiction to opium and opiates, could more easily thwart Japanese strategic ambitions which, as suggested, were partly dependent on a continuing, if not growing, market for opiates in North China and Manchuria.

Tokyo, however, need not have been overly concerned about the pace of opium suppression blocking the realization of its goals. The many domestic problems plaguing China slowed the consolidation of the Republican rule.[33] Indicative of the profound nature of the obstacles to effective governance was the mini-revolution in the summer of 1913, which resulted in nine southern provinces distancing themselves from Yuan's increasingly authoritarian rule. More important for present purposes was the fact that opium consumption meant big business in China, totaling at least US$150 million per year in provincial revenue. To produce crops other than opium would have been far less lucrative, although the cultivation of foodstuffs would have been far better for the health of the nation.[34]

While it might seem highly dramatic, it would not be wrong to conclude that the fate of the Republic hung in the balance by late 1915. Yuan had turned to an American, Frank J. Goodnow of Columbia University who was an expert on public administration and constitutional law, to draft a provisional constitution for China. Irrespective of the prospects for effective governance through more efficient administration, there was little likelihood that structural changes would provide a sense of stability any time soon.[35] In May, Japan issued the 21 Demands, a direct challenge to Chinese sovereignty. The immediate diplomatic responses from Washington and London to Japan's action expressed greater concern with the looming economic presence of Japan in Manchuria and the Shandong Peninsula than with the weaknesses of Yuan Shikai's government. Woodrow Wilson insisted that China and Japan maintain the inviolability of American rights under the Open Door policy. Great Britain, meanwhile, worried that its Far Eastern policy was in disarray and not up to meeting the Japanese challenge.[36] Yet, in the short term neither power knew what course of action would bring stability and security to China; they could only hope that Japan would act responsibly whatever its ultimate objectives.

In this chaotic situation, Yuan in effect abandoned his opium suppression policy and turned toward reliance on the opium business to produce the extensive revenues that could never be realized from the less lucrative salt gabelle. The device chosen was an opium combine, headquartered in Shanghai, under which all existing stocks of imported opium would be transferred to Beijing's control. The nominal purpose of the combine was to reduce the incidence of smuggling, increase badly needed revenues, and provide aged smokers with a reliable source of opium. In theory, the opium in question would replace the domestic variety whose cultivation was presumably in decline. To that end, authorities averred that the combine would turn out to be a means of achieving opium suppression by the end of March 1917, at which time the importation of opium from India was scheduled to cease. Not surprisingly, Sir John Jordan found the combine scheme wanting and predicted the resumption of a robust commerce in domestic opium beginning in April 1917.[37] Yuan Shikai, it must be said, had the good fortune to die in mid-1916 before it became clear that the combine scheme would actually exacerbate and prolong China's troubles with narcotics. As Kaku Sagataro had feared, opium did indeed seem to threaten the peace in Asia.

The US concern about the deleterious effects of opium on the Chinese people and the governance of China was born of a sense of *noblesse oblige* that accompanied merchants, missionaries, and diplomats in the late nineteenth century. Charles Denby, US Minister to China for 13 years starting in 1895, saw little immediate hope for a people and a land so in thrall to opium. 'The only thing that can be said for it', he observed, 'is that China-grown opium will eventually cut off the Indian trade, and at least enable the people to spend in their country the money which they devote to their own perdition.'[38] Denby's tough-minded analysis was not as much out of step with the progressive sentiments of the time as might be thought. That same relatively uncompromising reformist spirit had led the United States to advocate the cessation of opium smoking in the Philippines, the convening of the International Opium Commission in 1909 at Shanghai, and then the holding of a world antiopium parley at The Hague in 1911. Denby's blunt words can best be seen, therefore, as a barometer of sorts, measuring how hard Americans would have to work in China in order to realize their nation's own imperial ambitions.

In one respect, the issuance of the Twenty-One Demands had something of a galvanizing effect upon US policy in East Asia. It soon became more apparent that as pronounced as the differences were between British and American policies in China, they were differences in degree rather than in kind, particularly when it came to opium. Japan's advances into China greatly alarmed Methodist Episcopal Bishop James W. Bashford, who took it upon himself to speak for the Chinese republic, when he detailed for President Woodrow Wilson and Secretary of State William Jennings Bryan what he depicted as the Japanese threat to American religious and economic interests there. The British took no exception to increased American concern about China's fate and in fact welcomed the re-assertion of Open Door rights that the US Government believed to be at the heart of the Lansing-Ishii agreement of 1917.[39]

The increasingly worrisome situation in China ultimately forced the two powers to reconsider how they might address in concert, along with the Beijing Government, the matter of opium control. Merely ending the import trade from India was no panacea, what they needed was better coordination of their understanding of the enormity of opium's impact on China and how to deal with it. The British, thanks to Jordan, were ahead of the Americans in that effort; however, Washington's strong reaction to the opium combine scheme signaled a coming change in the US approach.

The very thing that colonial officials hard worked, so hard to eliminate in the Philippines, namely an opium monopoly, was about to become a reality in China. Diplomat John Van Antwerp MacMurray feared correctly that 'prominent members of the present government are interested [in it]'.[40] With the monopoly would inexorably come the breakdown of social and economic stability as domestic opium cultivation was sure to increase and the incidence of opium smoking would doubtless rise as well. Ambassador Walter Hines Page reported from London that Whitehall also opposed the prospective monopoly as contrary to the spirit of the 1912 Hague Opium Convention and the 1911 Sino-British antiopium accord. When the Chinese Government reached an agreement with the Shanghai Opium Combine in mid-1918 and actually took control of 1200 chests of opium, the Foreign Office joined the Department of State in issuing a protest.[41]

By the end of November 1918, pressure on the government moved President Xu Shichang, a former military protégé of Yuan Shikai, to destroy 80 tons of opium in a public burning in Shanghai, though it did not take place until January 1919. Support for this controversial endeavour, which would cost the government an untold amount of revenue, came from the several antiopium societies that were in the process of forming an umbrella organization, the International Anti-Opium Association (Peking).[42]

The very name of this agency hints at a deeper purpose that of promoting a reformed, modern China that would be more receptive to the financial and commercial interests of the Western powers. Among the most active members of the association were Elizabeth Washburn Wright, widow of Dr Hamilton Wright who had been one of the originators of the international antiopium movement; numerous foreigners living in Beijing, among whose numbers were Christian missionaries; and A.E. Blanco of the Chinese Maritime Customs who would later lead the Anti-Opium Information Bureau in Geneva as an ardent supporter of Generalissimo Chiang Kai-shek. Strong support for Xu's crusade against opium also came from the noted editor of the *Peking & Tientsin Times*, H.G.W. Woodhead; one of China's representatives at the antiopium conferences at The Hague, Dr Wu Liande; and, at Shanghai, Reverend A.L. Warnshuis of the International Missions Council. Such an array of prominent individuals, both foreign and Chinese, underlined the importance of antiopium activity not only for China's ability to govern itself but also for the protection of Western economic and strategic interests in the Republic.[43]

Sir John Jordan congratulated Xu, calling him 'an enlightened ruler'. Jordan, who had served in China since 1906, rejoiced that 'we have removed finally the stigma attaching in the popular mind to our connection with this traffic'. He added, giving voice to sentiments certainly shared in Washington, 'by keeping our own hands clean in the future we can continue to exercise pressure to prevent the exploitation of China by other countries'.[44] Implicit in his statement was the assumption that, unlike Japan, Britain and America were squarely on the side of the Chinese. The Japanese, who had dramatically increased narcotic production during World War I, were themselves confronted with vexing drug problems throughout their Chinese concessions, especially in the Kwantung Leased Territory where the illicit traffic in opiates festered out of control.[45] Meanwhile, US diplomats were encouraging Hsu's government to adopt additional antiopium laws, acting as though they originated entirely on his own initiative. Minister Paul Reinsch welcomed, for example, Xu's decision to ask foreign missionaries throughout China to report to local government officials the existence of opium cultivation whenever they discovered it. Also, Reinsch applauded the government's determination to place before the delegates at the Paris Peace Conference a suggestion that the agenda include the issue of opium control and narcotics suppression.[46] British representatives took the lead at Paris and persuaded a drafting committee to prepare what became Article 295 of the Treaty of Versailles, under which ratification of that pact meant acceptance of the 1912 Hague Opium Convention. Showing the fragility of the recent Anglo-American *démarche* over East Asian policy generally and antiopium policy specifically, the United States did not initially support the British effort at Paris. This was perhaps because, and it is not clear, the idea had not originated in the State Department.[47]

What was clear at the time, however, was that no international agreement would overcome the multitude of problems that opium posed for China, its security and governability. The Japanese-dominated trade in narcotics was becoming a matter of great concern in North China and Manchuria as were fissures within the republic itself. It is no small irony that until at least 1915 the narcotics that flowed from Japan into China, and then into that country's illicit trade, originated in Great Britain and, to a lesser extent, the United States.[48] Americans and British alike had for some time protested the presence of the vast quantities of Western morphine that were fueling the narcotics trade in China. They saved their strongest censure for Japan's apparent role in the business. E. Carleton Baker, US consul at Mukden, upon learning that the Japanese had apparently prohibited their nationals from participating in the trade, commented that 'their desire [is] to simply hoodwink the world and go through the motions of suppressing the evil'. While there is little doubt that the Japanese military administration in Tianjin was involved in or at least tolerated the narcotics trade, it is not as certain that government officials in Tokyo countenanced the activity. The press in Tokyo, for instance, was worried about the damage the traffic in narcotics would do to Japanese and prestige.[49]

The promise in February 1919 of the premier, Hara Kei (Takashi), to take action against the trade may have been genuine but not because of principled opposition to the business which the Japanese, like the British before them, viewed in free-market terms. Rather, the premier acted in deference to other, more immediate policy objectives. These included placing a statement on racial equality in the covenant of the proposed League of Nations charter. From his diplomatic post in Beijing, Sir John Jordan found some merit in Japan's professed commitment to undertake antinarcotic activity in northern China.[50]

China's internal political situation made control of the narcotics trade unlikely though. Failure to overcome the estrangement between northern and southern factions, located in Beijing and Guangzhou respectively, permitted the re-entry into Chinese politics of regional and provincial warlords. Maintaining their tenuous hold on power depended to some extent upon tax revenues obtained as a result of a resurgence of opium cultivation. The Zhili Clique, as the faction in Beijing was known, was powerless to halt opium growing in Fujian, in western regions of Henan, and in Shaanxi despite the fact that these all fell within its nominal jurisdiction. Not that the administrative structure in Guangzhou fared any better. Poppy cultivation flourished by 1920 in Hunan, Hubei, Guizhou, Yunnan, and Sichuan, all of which were theoretically under Guangzhou's control. In several of these regions local military figures, or *tuchuns*, commanded farmers to plant opium even at the expense of traditional food crops.[51]

Speaking for their governments, Jordan and Reinsch denounced these developments as a threat to 'China's national interests'.[52] They might as well have added 'and to our own'. There was little the two powers could do, however, other than issue a series of these empty protests as Republican China had embarked upon a highly divisive period that would last until 1928, during which regional warlords dominated the political landscape. The nascent spirit of revolutionary exultation had disappeared all too soon following the fall of the dynasty in 1911, leaving China in considerable turmoil and affording no comfort to those nations that hoped to strengthen their imperial identity on the East Asian stage. Even the Japanese were learning that the pursuit of these dreams of glory might have an extremely high cost.

Conclusion

The period from the end of the nineteenth century to 1920 saw four sets of players engaged in a set of rivalries in East Asia upon which the fate of empires depended. Opiates were at the heart of these rivalries throughout this period. The British sought to maintain their empire in South Asia and to extend their influence in China, and at various times used the trade in opium to finance their operations or as collateral in diplomatic strategies. The Americans sought to challenge the British imperial influence in Asia and, after establishing a colonial foothold in the Phillipines, focused efforts on a civilizing and commercial mission in China. Tackling the opium trade was at once a way of undermining the economics of British imperialism and of establishing good relations with the Chinese, who the Americans hoped would replace a need for opium with a desire for American goods. The Japanese sought to protect themselves against western interests by developing both economic and political might. Drugs trading was a means of doing both, as it allowed them to enter Chinese markets and politics and to extend their influence over both. Finally, the Chinese empire's demise was a story told in opium, as the Qing dynasty's inability to control consumption of and commerce in the drug fatally undermined its legitimacy as a government. The fragmentation of its successor state was in part due to its failure to assert a nationwide opium policy, and was certainly funded by warlords dealing in drugs.

The outcome of all of these imperial tergiversations was failure. Neither Great Britain nor the United States achieved their goals as China was too unstable for merchants, financiers, or missionaries to succeed in the long term. Japan, after successfully jostling for primacy in North China and Manchuria, would not soon derive the long anticipated benefits of its assertive China policy. As for the Chinese themselves, as the revolutionary decade came to a close it was impossible to discern exactly what remained of China's own identity, constructed as it long had been upon a foundation of unassailable imperial glory. Opium politics, in the first half of the twentieth century, proved to be a game that nobody won.

Notes

1. C.C. Chin, 'Beneficent Imperialists: American Women Missionaries in China at the Turn of the Twentieth Century' in *Diplomatic History*, 27, 2003, pp. 327–52.
2. M.W. Doyle, *Empires* (Ithaca: Cornell University Press 1986); D.C. Campbell, *Writing Security: United States Foreign Policy and the Politics of Identity* (Minneapolis: University of Minnesota Press 1992).
3. P.W. Fay, *The Opium War, 1840–1842* (Chapel Hill: The University of North Carolina Press 1975), pp. 44–9; J.K. Fairbank, *Trade and Diplomacy on the China Coast: The Opening of the Treaty Ports, 1842–1854* (Stanford: Stanford University Press 1969), pp. 68–71; Wu Wen-tsao, *The Chinese Opium Question in British Opinion and Action* (New York: Academy Press 1928), p. 19; Teng Ssu-yu, *The Taiping Rebellion and the Western Powers* (Oxford: Clarendon Press 1971), p. 28.
4. Fay, *The Opium War*, pp. 45–6; F. Wakeman, Jr., *The Fall of Imperial China* (New York: Free Press 1975), p. 128; Fairbank, *Trade and Diplomacy*, pp. 76–8 and 240–3.
5. Jen Yu-wen, *The Taiping Revolutionary Movement* (New Haven: Yale University Press 1973), pp. 444–5; Teng, *The Taiping Rebellion*, pp. 284–316; H.B. Morse, *The International Relations of the Chinese Empire*, vol. 2: *The Period of Submission, 1861–1893* (London: Longmans 1918), pp. 90–111.
6. M. Mancall, *China at the Center* (New York: Free Press 1984), pp. 139–41, 145–60 and 184–5; Wakeman, *The Fall of Imperial China*, pp. 185–6 and 189–95; J. Spence, 'Opium Smoking in Ch'ing China' in F. Wakeman, Jr. and C. Grant, (eds), *Conflict and Control in Late Imperial China* (Berkeley and Los Angeles: University of California Press 1975), pp. 151–3.
7. Teng Ssu-yu and J.K. Fairbank (eds), *China's Response to the West: A Documentary Survey, 1839–1923* (Cambridge: Harvard University Press 1965), p. 100.
8. H.B. Morse, *The Trade and Administration of China*, 3d rev. ed. (New York: 1967), pp. 368–9.
9. Quoted in Wu, *The Chinese Opium Question*, p. 135.
10. Teng and Fairbank, *China's Response to the West*, p. 170.

11. U.S. Department of State, *Papers Relating to the Foreign Relations of the United States, 1906*, Pt. 1 (Washington, DC: 1909), pp. 357–8; hereafter *FRUS [year]*.
12. A.H. Smith, *The Uplift of China* (New York: 1907), p. 215 and n. 1; Philippine Commission, Opium Investigation Committee, *Use of Opium and Traffic Therein: Message from the President of the United States, Transmitting the Report of the Committee Appointed by the Philippine Commission to Investigate the Use of Opium and Traffic Therein*, 59 Cong., 1 sess., S. Doc. 265 (Washington, DC: 1906), pp. 52–5; Morse, *The International Relations of the Chinese Empire*, vol. 2: *The Period of Subjection, 1894–1911* (London: Longmans 1918), p. 437, hereafter OIC.
13. M.E. Cameron, *The Reform Movement in China, 1898–1912* (Stanford: Stanford University Press 1931), pp. 143–5; *China Centenary Missionary Conference Records, Report of the Great Conference Held at Shanghai, April 5th to May 8th, 1907* (New York: 1907), pp. 387–90.
14. Morse, *The Trade and Administration of China*, pp. 373–78; Morse, *The International Relations of the Chinese Empire*, vol. 3, pp. 437–8; Great Britain, Foreign Office, *The Opium Trade, 1910–1941*, FO 415: Correspondence Respecting Opium, Public Record Office, 6 vols (Wilmington, DE: 1974), vol. 1, III, January–June 1911, no. 183, hereafter FO; FO, *The Opium Trade*, vol. 1, I, January–June 1910, nos. 21 and 25.
15. W. LaFeber, *The Clash: A History of U.S.-Japan Relations* (New York: Norton 1997), pp. 65–104.
16. A.H. Taylor, *American Diplomacy and the Narcotics Traffic, 1900–1939: A Study in International Humanitarian Reform* (Durham: Duke University Press 1969), pp. 53–8 and 61–2; *FRUS, 1908*, pp. 100–2.
17. Taylor, *American Diplomacy and the Narcotics Traffic*, pp. 65–80.
18. FO, *The Opium Trade*, vol. 1, II, July–December 1910, no. 63 and enclosure.
19. LaFeber, *The Clash*, pp. 93–98.
20. FO, *The Opium Trade*, vol. 1, II, July–December 1910, no. 80; IV, July–December 1911, nos. 117, 180; Taylor, *American Diplomacy and the Narcotics Traffic*, pp. 87–9 and 93.
21. Wakeman, *The Fall of Imperial China*, pp. 97–120.
22. My thanks to Alex McKay for providing this important information.
23. FO, *The Opium Trade*, vol. 2, V, January–June 1912, no. 81.
24. I.H. Nish, *The Anglo-Japanese Alliance: The Diplomacy of Two Island Empires, 1894–1907* (London: Athlone 1966), pp. 142, 365, 369, and 372–77.
25. For a contrasting interpretation of Japan's role in the antiopium movement, see J. M. Jennings, *The Opium Empire: Japanese Imperialism and Drug Trafficking in Asia, 1895–1945* (Westport, CT: Praeger 1997), pp. 62–64.
26. I.H. Nish, *Alliance in Decline: A Study in Anglo-Japanese Relations, 1908–1923* (London: Athlone 1972), pp. 81–2, 99–105, and 392–3; P. Lowe, *Great Britain and Japan, 1911–1915: A Study of British Far Eastern Policy* (London: Macmillan 1969), pp. 58–88 and 99–117.
27. OIC, *Use of Opium and Traffic Therein*, pp. 24–8, 68, and 73; Jennings, *The Opium Empire*, pp. 22–23.
28. E.P. Young, *The Presidency of Yuan Shih- k'ai: Liberalism and Dictatorship in Early Republican China* (Ann Arbor: University of Michigan Press 1977), pp. 52–5; S.R. MacKinnon, *Power and Politics in Late Imperial China: Yuan*

Shih-k'ai in Beijing and Tianjin, 1901–1908 (Berkeley: University of California Press 1980), pp. 35–45 and 138–79.

29. FO, *The Opium Trade,* vol. 2, VI, July–December 1912, no. 113 and enclosure 1.
30. Ibid., V, January–June 1912, no. 89; VI, July–December 1912, no. 124 (quotation).
31. Ibid., VI, July–December 1912, nos. 1, 59 and enclosure 144; V, January–June 1912, no. 64 (quotation).
32. Young, *The Presidency of Yuan Shih-k'ai*, pp. 123–8, 134, and 180; Lowe, *Great Britain and Japan*, pp. 103, 127–9, and 132–43.
33. R. Bin Wong, 'Opium and Modern Chinese State-Making', in T. Brook and B.T. Wakabayashi (eds), *Opium Regimes. China, Britain, and Japan, 1839–1952* (Berkeley: University of California Press 2000) passim.
34. Young, *The Presidency of Yuan Shih-k'ai*, pp. 120–33; Jerome Ch'en, *Yuan Shih-k'ai: Brutus Assumes the Purple* (Stanford: Stanford University Press 1961), pp. 163–8; J.E. Sheridan, *China in Disintegration: The Republican Era in Chinese History, 1912–1949* (New York: Free Press 1975), pp. 47–54; *FRUS, 1913*, pp. 124, 127, and 219.
35. N.H. Pugach, *Paul S. Reinsch: Open Door Diplomat In Action* (Millwood, NY: KTO Press 1979), pp. 54 and 124–7; *FRUS, 1915*, pp. 46–58.
36. Mancall, *China at the Center*, pp. 201–2; G.P. Gooch and H. Temperly (eds), *British Documents on the Origins of the War, 1898–1914*, vol. 10: *The Last Years of* Peace, pt. 2 (London: HMSO 1938), enclosure in no. 497 and Appendix 1; Nish, *Alliance in Decline*, pp. 105 and 155–8; *FRUS, 1915*, pp. 84, 93–5, 111, 146, and 159–70; A.S. Link *et al.* (eds), *The Papers of Woodrow Wilson*, vol. 32: *January 1–April 16, 1915* (Princeton: Princeton University Press 1980), pp. 139, 322–4, 520–1, and 531; Sheridan, *China in Disintegration*, pp. 50–6.
37. Young, *The Presidency of Yuan Shih-k'ai*, p. 241; FO, *The Opium Trade*, vol. 3, IX, January–November 1915, nos. 21–22, 38 and enclosure 5.
38. C. Denby, *China and Her People*, 2 vols (Boston: Page 1905), vol. 2, p. 21.
39. A.S. Link *et al.*, eds, *The Papers of Woodrow Wilson*, vol. 33: *April 17–July 21, 1915* (Princeton: Princeton University Press 1915), pp. 23–6 and 31–7 (quotation, 35); *FRUS, 1917*, pp. 260–6 and 273.
40. J.V.A. MacMurray to the Department of State, 26 August 1918, General Records of the Department of State, Decimal Files, Record Group 59 893.114/74, National Archives Microfilm Publication, M329, roll 113, National Archives II, Suitland, MD.
41. Ibid.; FO, *The Opium Trade*, vol. 4, XII, 1918, nos. 16–7 and 19.
42. P.S. Reinsch to DS, 25 November 1918, RG 59, 893.14/185, NAMP, M329, roll 113; *North China Herald*, 28 September 1918; International Anti-Opium Association (Peking), *The War Against Opium* (Tientsin: 1922), pp. i–ii.
43. Ibid., pp. ii–iii; T. Sammons to Reinsch, 18 January 1919, RG 59, 893.114/206, NAMP, M329, roll 114.
44. FO, *The Opium Trade*, vol. 4, XIV, 1919, no. 13.
45. Jennings, *The Opium Empire*, passim.
46. Reinsch to DS, 24 February 1919, RG 59 893.114/219, NAMP, M329, roll 114, NA.
47. Taylor, *American Diplomacy and the Narcotics Traffic*, pp. 142–4; *FRUS, 1919*, vol. 4: *The Paris Peace Conference*, pp. 552–3, 567–9, and 595.

48. Jennings, *The Opium Empire*, pp. 65–6.
49. Baker to DS, 27 July 1918, RG 59 893.114/179, NAMP, M329, roll 113; W.R. Peck to Reinsch, 21 November 1918, and Peck to DS 6 July 1919, W.R. Peck Papers, box 3, Hoover Institution on War, Revolution, and Peace, Palo Alto, CA.
50. Nish, *Alliance in Decline*, pp. 258, 265; G.W. Egerton, *Great Britain and the Creation of the League of Nations: Strategy, Politics, and International Organization, 1919–1945* (Chapel Hill: University of North Carolina Press 1978), pp. 164 and 168; FO, *The Opium Trade* vol. 4, XIII, 1919, no. 16; XV, January–June 1921, nos. 6, 78 and enclosures, 79 and enclosures.
51. G.C. Hanson to Reinsch, 12 October 1918, RG 59, 893.114/192; and Sammons to Reinsch, 16 January 1919, RG 59, 893.114/202, NAMP, M329, roll113, NA; FO, *The Opium Trade*, vol. 4, IV, 1920, no. 20 and enclosure.
52. Ibid., vol. 4, XIII, 1919, no. 16 and enclosure; Reinsch to DS, 29 April 1919, RG 59, 893.114/231, NAMP, M329, roll 113, NA.

12
'Wolf by the Ears': The Dilemmas of Imperial Opium Policymaking in the Twentieth Century

William B. McAllister

Introduction

> We have a wolf by the ears, and we can neither hold him nor safely let him go. Thomas Jefferson to John Holmes, 22 April 1820.

Though Jefferson's famous statement referred to slavery, the imagery would no doubt have resonated with twentieth-century officials who attempted to forge opium policies that accommodated the disparate interests of imperial regimes. Colonial and metropolitan administrators repeatedly foundered on the insoluble 'damned if you do, damned if you don't' nature of the opium question. This chapter focuses on the British intra-imperial experience, thereby elucidating certain common patterns that illuminate similar circumstances faced by French, Dutch, Portuguese, and Japanese colonial regimes. Every imperial government involved in the opium trade became, to borrow from Lincoln, a house divided against itself as each wrestled with how to accommodate the call of their better angels. This chapter will examine the ways in which the opium issue generated and sustained these internal divisions in imperial governments in the first half of the twentieth century.

The authors of this book and others have detailed many aspects of the early history of international drug exchange and its concomitant geopolitical effects, obviating the need for detailed discussion in this chapter.[1] In brief, China objected to rapidly increasing amounts of opium entering the country from the late eighteenth century onwards. The principal suppliers and shippers operated under the British and the Portuguese flags. The Opium Wars of the mid-nineteenth century exacerbated the problem within China, resulting in large increases in opium imports, domestic production, and consumption. By the last quarter of the nineteenth century, several constituencies in western nations, particularly in Great Britain, objected to the India-to-China opium trade. Pressure increased to impose some type of control on the commerce and this ultimately resulted in a 1906 agreement between Britain, India, and China. This was based on a novel form of reciprocity whereby the Indian Government agreed to curb opium exports to China, while Beijing committed to reduce domestic agricultural production. A treaty signed at the Hague in 1912 created the statutory basis for control in international law. The peace treaties following World War I and the Covenant of the League of Nations consolidated the control impetus.[2] By the early 1920s, governments generally felt committed to moving toward placing substantive restrictions on the international traffic in addictive substances.

Moreover, in the later nineteenth century the desire to reform problematic 'opium farm' sales-distribution systems resulted in the imposition of opium monopolies administered by colonial government officials. By the early twentieth century, however, those arrangements had come under fire from critics claiming that colonial governments had become reliant on the profit from sales of dangerous substances to Asians. After World War I, imperial governments featuring opium monopolies experienced difficulties in justifying and defending their positions, not least of all to themselves. Great Britain serves as a useful case study for intra-imperial drug relations because the difficulties of policy formation and implementation mirrored that of other governments with imperial interests. Moreover, the British example serves as a proxy for the international nexus, a sort of world-in-miniature. All the problems experienced in global negotiations had their counterparts in British intra-imperial relations.[3]

Britain and its opium factions

One cohort within the imperial structure generally favored substantial regulatory measures. Led by the Home Office (HO) and the Foreign Office (FO), these pro-control advocates adopted the perspective of the internationally oriented metropole. The HO wished to prevent domestic drug abuse and favored regulation of the pharmaceutical industry and medical profession to achieve that end. At the same time, the HO understood the value of British drug firms to the domestic economy and their ability to earn foreign exchange via the export market. The FO desired to promote internal and international comity on the increasingly thorny drug issue. FO officials usually favored a balanced approach in the hope that all parties would compromise and accept enough regulation to curb the most notable excesses of the drugs traffic while avoiding measures that would overly restrict international commerce in general. They had in mind a level-playing field for British pharmaceutical exports and also had an eye on political issues arising from the potential for international obligations to conflict with national sovereignty.

The colonial officials responsible for day-to-day management of imperial possessions generally opposed international drug control initiatives. The India Office[4] (IO) and the Colonial Office (CO) comprised the principal elements of the opposition. Viewing the situation from a parochial perspective, imperial administrators stressed the burdens that control would place on colonial authorities and the improbability that such measures would succeed. They emphasized that opiate use was a long-standing phenomenon among significant segments of indigenous populations and that attempts to uproot the practice would cause more problems than they solved. They preferred to manage the problem locally rather than commit to prescriptive prohibitions that were politically popular in the metropole, but problematic to enact in the field. Although colonial officials increasingly conceded the odious nature of the opium business, they insisted that the cure would be worse than the disease.

The Dominions Office[5] (DO) and the Board of Trade (BOT) played less central but nevertheless important roles as moderating influences most interested in promoting free trade. The DO often served as a voice for the consumers of medicinal products in those territories that possessed neither production capacity nor pharmaceutical manufacturing capability but which still required medicines at reasonable prices. Though recognizing the need for organized control the Dominions acted as a brake on overly restrictive initiatives by insisting on sufficient licit agricultural production and pharmaceutical manufacture to ensure affordable opiate-analgesics. The BOT regarded drugs purely as a commodity, eschewing the moral overtones that clouded the debate in other quarters. Trade officials were concerned that overly restrictive drug-control treaties might interfere with other commercial relationships, especially in the realms of agriculture and chemical manufactures. The BOT generally attempted to blunt onerous control initiatives in order to avoid retaliation against other British products.

Two British officials embodied the differing drug-control positions that were embedded in the Empire. As the chief representative for India, the irascible and cantankerous Sir John Campbell took a combative approach toward internal as well as external attempts to compromise the interests of the colonial government there. The redoubtable Sir Malcolm Delevingne of the Home Office served as the linchpin of British-Imperial policy, and during the interwar years he played an important role in the debates concerning domestic, imperial, and international drug policy. Delevingne suffered the burdens of trying to craft a policy that all of the parties outlined here, and of course all of the interested nations, could accept and work to implement. That he achieved only limited success should come as no surprise given the conflicting positions with which he had to contend.[6]

Dissent in the ranks, 1920–1925

During the early 1920s, various elements of the British/imperial government participated actively in forging the international control regime. However, these often worked at cross-purposes. Most importantly and uniquely, the colonial government in India secured separate representation from Great Britain in the international drug-control apparatus. The Indian Government sent its own voting member to the Opium Advisory Committee, Indian experts served independently on other key League committees that dealt with the opium question, the Indian Government enjoyed separate representation at the two plenipotentiary meetings of 1924–1925, and the Permanent Central Opium Board included an Indian member from its inception in 1928.[7] No other British possession and no other imperial power received this privilege. On the one hand, this dual representation indicates the magnitude of British influence as well as a clear recognition of India as a major power in its own right. However, this Britannic Gemini also signified British colonial, especially Indian, administrative independence and indicated the potential for problematic intra-imperial relations. As the international drug-control apparatus became operational during the 1920s, the Indian Government and other British colonial administrations quickly demonstrated a propensity to steer a separate course from that promoted by Whitehall and the HO.

A constellation of related issues bedeviled drug diplomats, all of which highlighted rifts within the imperial camp. Indian-colonial experts defended the medical efficacy of opium, thereby supporting a key basis for the legitimacy of colonial opium monopolies. They coined the term 'quasi-medicinal use' to describe the complex of traditional opiate consumption patterns, including, in addition to the ubiquitous smoking habit, use for prophylactic, analgesic, anti-diarrhoeal, anti-fever, religious, and social purposes. Metropolitan British authorities were much less sanguine about the medicinal value of raw or prepared opium, conceding their use as a temporary necessity until what they viewed as more modern health measures could be initiated. Consequently, the two parties calculated very differently how much opium was necessary for medicinal and quasi-medicinal purposes which in turn affected international negotiations aimed at reducing supplies to 'legitimate need'.[8]

Differing assessments of medicinal need caused metropolitan and colonial authorities to form contrasting views about the international trade in opium products. The IO favored the perennially profitable export of Indian opium to fulfill the 'legitimate' needs of colonial monopolies, as calculated by monopoly officials. Officials from other colonies, especially British territories, preferred importing opium from India because of its consistency and stable cost. The HO and FO viewed this self-reinforcing system as less and less defensible over time, especially as international political opprobrium increased. Metropolitan and colonial officials differed on the related issue of the illicit traffic. The CO, based on multitudinous reports from officials posted to Asia, feared that imposing substantive controls over distribution would cause an increase in smuggling, drive users to illicit sources of supply, and degrade enforcement agencies through exhaustion and corruption. The FO and HO faced rising international pressure to diminish the inter-colonial trade and insisted that the only viable solution was to reduce the opium question to the dimensions of a police-enforcement issue.

Difficulties with and within China highlighted the central aspects of the problem. Everyone understood that massive Chinese consumption and substantial domestic production overshadowed all attempts to curb the international opium trade. Colonial officials argued that there was little point in restricting internal use and production for export. Doing so would simply transfer profits to other colonial administrations and to independent states willing to wink at the trade: Persia and Turkey were the most conspicuous candidates in the 1920s, Japan in the 1930s. Moreover, restricting British-produced supplies might simply stimulate Chinese domestic cultivation and illicit trafficking.[9]

Consequently, as international drug control negotiations progressed during the first half of the 1920s, the IO and CO objected to the many initiatives designed to restrict or eliminate opium smoking and quasi-medicinal opium use. Representatives of the colonies eschewed compromises urged upon them by their compatriots in London. The Indian Government's representative, Sir John Campbell, received his instructions from the IO rather than the FO and often adopted a narrow interpretation of his latitude to make concessions. During tense intra-imperial negotiations in mid-1924, the IO even claimed to be responsible to the League of Nations rather than His Majesty's Government.[10]

As international negotiations came to a head in 1924–1925, the entire drug control agenda was driven in large part by the need to accommodate recalcitrant colonial administrations. Sir Malcolm Delevingne of the HO, dean of drug diplomats, wished to promote an omnibus drug-control treaty that would regulate all aspects of the trade. Delevingne understood, however, that the colonial powers would not agree to substantive limitations on opium smoking and opium monopolies. Indeed, it was elements of his own government, the IO and CO, which provided some of the fiercest opposition to his plans. Delevingne and other control advocates capitulated by authorizing a smaller conference at which the colonial powers would discuss the multifaceted Far Eastern opium problems. Although the ostensible goal was to suppress opium smoking and quasi-medical use in South and East Asia, few believed that anything significant would come of the meeting because none of the opium-imperial powers favored sweeping reforms. Owing to the immense intra-imperial dissent as well as inter-governmental contention that the opium/opium smoking dilemma caused, control advocates judged that the only way to make progress on other matters was to separate that issue from the larger international agenda. The goal of the much broader second conference was to forge a general drug-control regime for the global trade in addictive substances. Yet that task remained impossible without meaningful progress on the opium question in Asia. The colonial powers' schizophrenic approach to drug control, therefore, resulted in two separate but intimately intertwined League-sponsored plenipotentiary conventions taking place simultaneously during 1924–1925. In this way the very structure of the international drug-control regime was driven by intra-imperial conflict.[11]

This climax of early drug-diplomatic negotiations caused the long-standing cracks in British intra-imperial policy to fracture openly in mid-1924. The FO and HO believed it politically expedient to promote international accord by acceding to opium suppression, even if the date for complete proscription was set many years in the future. Colonial governments, however, believed it impossible to suppress opium smoking. The British authorities in Hong Kong admitted that they could function without the revenue derived from opium sales, but they also maintained that enforcing such a ban would entail jailing 20 per cent of the population. The Straits Settlements calculated that, with great effort, they could probably do without the opium revenue in 10 years. Otherwise their position mirrored that of Hong Kong. In the Malay States the financial situation was even worse. The government depended on opium revenue to operate many of the colony's medical, educational, and social services. Local officials also feared disorder if a ban was enforced immediately. A previous attempt to register smokers and restrict imports had caused widespread smuggling, hoarding, and speculation to the extent that after a few months the government had abandoned the experiment. The IO threatened open rebellion, denying any higher power the right to determine its opium policy. India wanted to preserve for colonies the right to decide about regulating opium smoking and insisted on its prerogative to continue opium exports to any government that legally requested them, even to territories suspected of fostering illicit diversion. The matter reached an impasse so severe that all parties appealed to the Cabinet, although that body proved no more able to make a decision. Some believed it was time to eschew the goal of supply control as impossible to achieve, but most policymakers thought the political cost for admitting publicly that Britain could not fulfill its obligations would be too high. When the Labour Government suddenly fell in October 1924, Delevingne and his colleagues were left with no policy only weeks before the negotiations began.[12]

At that point, Delevinge took the extraordinary step of attempting to delay the beginning of the two conferences. His superiors and League officials refused to sanction this on the grounds that any postponement would cause a firestorm of criticism among those who doubted the motives of the British government and the efficacy of the League. Yet Delevingne's desperate move illustrates the internal difficulties experienced by all the imperial powers; he desired a deferral not because he required more time to build consensus among other delegations but rather because he could not corral the disparate elements of *his own* government.[13]

Hamstrung by this fundamental rift between periphery and metropole, Malcolm Delevingne foundered amidst the competing agendas of various British representatives during the 1924–1925 negotiations. The HO and FO attempted to forge compromise and were willing to entertain restrictions on opium and other drugs. Delevingne also kept in mind the reasonable requests of the Dominions for access to medicinals, and he carefully tried to maintain a level playing field for British pharmaceutical products in order to satisfy the BOT constituency. The IO spearheaded the Asian colonial administrations' objections to any substantive restrictions on their authority to regulate, or not regulate, the opium trade. The debates became so heated that League officials, fearful that the negotiations might collapse, called in Secretary-General Eric Drummond in an attempt to foster amity among the fractious delegations. Undeterred, IO representative Campbell engaged in shouting matches with the head of the American delegation. This turn of events shocked Drummond who appealed to London to control Campbell.

At the same time, other governments with significant colonial interests exhibited symptoms of this same bifurcation in policy. The French, Dutch, Portuguese, and Japanese displayed a reticence to impose significant restrictions on the Far Eastern trade in opium, while at the same time desiring to maintain at least the appearance of international cooperation.[14] During a Christmas recess called to forestall collapse of the negotiations, all the major colonial powers held emergency meetings in their respective capitals, signaling the depth of their internal divisions. Although most deliberations took place behind closed doors, the British Government again could not avoid airing its laundry publicly; Campbell wrote a letter to the *Times* in an attempt to garner support for the IO position. As a consequence of the uproar, most imperial governments appointed high-ranking political representatives to head their delegations as the sessions reconvened in January 1925. In each case the pro-colonial factions carried the day, and the two treaties produced by the dual conferences placed few substantive restrictions on colonial administrations' prerogatives with regard to the opium trade.[15]

Formation of the Interdepartmental Opium Committee, 1925–1927

For all colonial powers the outcome of the 1924–1925 negotiations represented a public-relations debacle. The East Asian imperial states left the impression that they were overly solicitous toward a trade that suffered increasing global disrepute. Despite the legitimate, and indeed prescient, objections of those pointing out that restriction would cause more problems than it solved, moral opprobrium became increasingly attached to those purveying opium to their subjects or across borders.

In a significant intra-imperial development the storm of controversy surrounding the opium issue caused the previously resolute Indian Government to modify its policy. In 1925, India halted opium exports to territories that appeared to foster the illicit traffic, even when those governments produced legal import certifications. The following year, the government announced that India would decrease and eventually curtail exports of opium for quasi-medicinal use. Taken together, these changes undermined key justifications for continuing the status quo by implying the illegitimacy of quasi-medicinal use and by suggesting that the costs of continued exports outweighed the benefits.[16]

Yet India's effort to demonstrate good intent to the international community only exacerbated Britain's intra-imperial problems. The CO continued to support opium monopolies and quasi-medicinal use, maintaining their position that suppression would lead to greater smuggling and enforcement woes. After India's reversal Colonial Office, officials moved to re-organize the intra-imperial power structure by attempting to supplant the HO as the centre of the British drug control apparatus. When that effort failed the CO next appealed for a status equivalent to the semi-independent IO by demanding separate representation in the League drug-control organizations. After the central government rebuffed that initiative, the CO announced that if they could not be supplied from Indian stocks, colonial administrations would purchase opium supplies for their monopolies elsewhere. A disconcerting bidding war ensued featuring negotiations between Crown Colonies and jurisdictions known as illicit trafficking havens, most notably Persia.[17]

Many within governmental circles recognized that it did not serve British interests to continue displaying such policy disputes in public. To foster a more coordinated approach, all parties agreed in 1927 to create an Interdepartmental Opium Committee (IOC) that would negotiate intra-imperial disagreements. Representatives of the HO, FO, IO, and CO comprised the core of the IOC, with the DO, Burma Office, BOT, military representatives, and others attending when the agenda encompassed their interests. For the next two decades, the IOC met on an as-needed basis[18] and generally acted as a damper on the internecine debates concerning opium that bedeviled all colonial powers. Notably, the other government most deeply riven by opium policy, Japan, also formed an inter-governmental opium committee in an attempt, ultimately unsuccessful, to assert civilian control over an increasingly chaotic situation in Northeast Asia.[19]

Changing places, 1928–1938

The focus of drug control efforts shifted in the later 1920s and early 1930s in large measure because of conflicts and constraints inherent in imperial opium policy. Few saw any prospect of headway on the Asian opium front, a view reinforced by another treaty negotiated in 1931 that produced inconsequential results. As a result of the stalemate on opium, Delevingne and others turned their attention to manufactured drugs. The important Manufacturing Limitation Convention emerged from that effort in 1931 and formed the basis for regulation of heroin, morphine, codeine, and other opiates. In 1936 the League sponsored another plenipotentiary meeting at which governments negotiated an anti-trafficking treaty. Few states ratified it, however, largely because those interested in making serious enforcement efforts were already doing so. In the later 1930s, governments even considered a quixotic plan to curtail agricultural production of opium. Yet all those efforts essentially avoided the principal issue, that of the burgeoning opium problem in East Asia.[20]

Although the opium situation in and around China had continued to be problematic throughout the 1920s, the Japanese occupation of Manchukuo in 1931 engendered a massive upswing in trafficking and abuse. The Manchukuo Government imported large amounts of opium from rogue states such as Persia and few believed Japanese explanations about the monopoly's intent to reduce individual users' consumption gradually. Accusations abounded that the Japanese military was also involved in, and profited from, drug trafficking. Opium trafficking and illicit trading in manufactured drugs such as morphine and heroin swelled, and new supplies not only fomented debility in China, but also threatened to inundate colonial territories and the west coast of North America. The FO/HO and IO/CO alliances recognized the threat to their interests, as did Dominions such as Canada, Australia, and New Zealand. Despite considerable negative publicity, the Japanese civilian government seemed unable or unwilling to curb the illicit activity.[21]

Britain's larger imperial interests, however, precluded a direct challenge to Tokyo. The belligerence of Japanese foreign policy, backed by considerable military power, caused Whitehall to temporize. France, the Netherlands, and Portugal also feared Japanese power but their lukewarm commitment to drug control enabled them to sidestep the issue more easily.[22] In any case, growing security concerns in Europe and an imploding world economy made it impossible to adopt anything but a defensive policy in the Pacific rim. The British Government recognized that continuing opium trafficking destabilized the entire region, and colonial officials reiterated that in such a supply-rich environment any attempt to eliminate monopolies would result in smuggling on a massive scale. The desire to preserve an increasingly fragile imperial regime dictated an appeasement strategy that only exacerbated the problem; British representatives soft-pedaled the opium issue at League meetings, hoping in vain to entice more cooperation out of Tokyo.[23]

In this environment, intra-imperial opium relations during the 1930s became somewhat more quiescent. All segments of the British Government wished to distance themselves from the Japanese example and to avoid culpability for contributing to the Chinese situation. The CO and its constituent administrations shifted from strident defense of their prerogatives to recognition of the fundamentally problematic aspects of official participation in the opium business. They increasingly referred to their opium monopolies as a necessary evil, and gone were the days when colonial administrations trumpeted monopolies as a progressive reform measure. The Indian Government adopted a more moderate strategy, gradually curtailing opium exports to colonial monopolies and eschewing the self-justifying rhetoric of earlier decades. The FO and HO concentrated more on attempts to cajole cooperation from Tokyo. In light of Japanese practice, British transgressions seemed less serious and the need to placate Japan caused London to demand less from its own colonial representatives.[24]

Contentious issues still arose: for example, although not admitted publicly, British trade officials advocated allowing British shipping firms to continue participating in the Persia–Manchukuo trade. The only British trading house still operating in Persia required receipts from the opium trade to stay afloat and the BOT argued that others, particularly the Japanese, would simply take over the profitable trade if British firms voluntarily withdrew.[25] Nevertheless, as the 1930s progressed, IOC meetings became less contentious as all players struggled with how to manage an increasingly problematic status quo.

World War II and after

By the late 1930s the threat of impending war caused British officials to contemplate increasing India's capacity to produce and export opium. If hostilities broke out the demand for morphine, codeine and other opium-based medicines would skyrocket. India was the only secure source of supply for Great Britain and the United States. American officials began stockpiling Indian opium and even contemplated growing opium on US soil in order to ensure adequate raw material supplies. Under these circumstances, intra-imperial opium relations took on a different cast and colonial monopoly needs competed with more 'legitimate' requirements and the bureaucrats concentrated on how best to distribute an increasingly valuable strategic material.[26]

Although the Japanese temporarily solved the colonial powers' public relations problem by occupying British, French and Dutch possessions at the beginning of the Pacific War, post-war planning soon returned the issue to the forefront. By late 1942, it was apparent that the Americans had stemmed the Japanese offensive and would likely liberate former colonial territories as part of their advance. This scenario presented a dilemma for the European imperial powers because American public opinion opposed the reinstitution of colonial opium monopolies. Harry Anslinger, Commissioner of the United States Bureau of Narcotics and staunch foe of opium monopolies and legalized opium smoking, engineered a masterful campaign of intimidation to force the issue. He sponsored a series of informal meetings and intimated that the United States would suppress opium monopolies in those areas it occupied if the British, Dutch, and French did not do so of their own accord. The Commissioner even utilized the Dominions relationship to create pressure from within by including Canadian and New Zealand representatives in the talks. Anslinger eventually convinced the State Department to fall in line with this confrontational approach, despite reservations that forcing the hand of the Allies on the drug question might endanger cooperation on larger issues.[27]

Fearful that the timely return of colonial territories to British authority might be endangered, even the CO and the Burma Office (BO) recognized the necessity to come to grips with the situation. In a momentous IOC meeting on 7 September 1943, all elements of the British Government concurred that announcing a policy of suppression of opium monopolies and opium smoking before it was imposed from the outside was the only course of action available. Indeed, in order to salvage the public relations benefit that would accrue to a voluntary renunciation, the British and the Dutch coordinated efforts to make simultaneous announcements about the new policy.[28] This dénouement represents not so much a triumph of the metropole as an indicator of the demise of British colonial power. Opium had proven to be a crucial contributor in building the empire and now the inability of the British to drive their own opium policy served as a symbol of imperial decline.

After 1945 the warnings of colonial administrators proved largely correct. The Dutch soon lost their East Indies colony, not least of all because opium revenue flowed to those fighting for independence. The French ability to control Indochina waned as other players took over the opium traffic.[29] Try as they might, officials in Hong Kong, Malaya, Singapore, and other British colonies could not stem smuggling and illicit use.[30] Nevertheless, after a half-century of internal disagreement all segments of the British Government had adopted a similar policy with regard to the opium question. Indeed, intra-imperial conflict receded sufficiently that no one felt the need to convene meetings of the Interdepartmental Opium Committee after 1946. That this uniformity came about largely as a result of external pressure indicates how opium serves as a barometer of imperial independence; as control over opium policy dwindled, so did the Empire.

The Wolf: Opium as a biological imperial power

During the twentieth century, opium-related foreign policy was made up of a series of compromises and power negotiations that reflected the inner tensions of the imperial project. Opium serves as an illustrative example of the clash of interests between metropole and periphery on a variety of issues. The colonial powers had at least as much trouble managing internal disputes as dealing with external affairs. Elements within each government proved as susceptible to opium addiction as individuals. The policy shifts of the twentieth century may, in fact, represent only the first phase of a cathartic withdrawal process, the outcome of which is still uncertain.

By 1945 the political economy of the Asian opium question had come full circle. After several centuries of utilizing the trade to promote imperial aggrandisement, the European colonial powers found themselves on the other side of the equation. British, French, Dutch, and Portuguese imperial interests fell prey to a more aggressive Japan that used opium to enhance its power. The Japanese, however, entered into a devil's bargain; opium-related problems on the Asian continent fomented chaos that contributed significantly to Japan's downfall. Opium played a prominent role in both the creation and the dissolution of all the great East Asian empires. Much like Jefferson's wolf, it is perhaps appropriate to interpret opium as an actor in its own right. Rather than simply an inert substance, opium might be seen over the last three or four centuries as a sort of independent biological imperial agent. In recent decades its worldwide ubiquity only confirms the wolf's power; opium appears to have bested all its human contenders.

Notes

1. D.E. Owen, *British Opium Policy in India and China* (New Haven: Yale University Press 1934); J. Walvin, *Fruits of Empire* (Basingstoke: Macmillan 1997); C.C. Stelle, *Americans and the China Opium Trade* (New York: Arno Press 1981); J.R. Rush, *Opium to Java: Revenue Farming and Chinese Enterprise in Colonial Indonesia, 1860–1910* (Ithaca, NY: Cornell University Press 1990); C. Trocki, *Opium and Empire: Chinese Society in colonial Singapore,*

1800–1910 (London: Routledge 1999); T. Brook and B.T. Wakabayashi (eds), *Opium Regimes: China, Britain, and Japan, 1839–1952* (Berkeley: University of California Press 2000); J.D. Spence, 'Opium' in J. D. Spence (ed.), *Chinese Roundabout: Essays in History and Culture* (New York and London: W.W. Norton 1992); P.W. Fay, *The Opium War, 1840–42* (Chapel Hill: University of North Carolina Press 1975); J.K. Polachek, *The Inner Opium War* (Cambridge, MA: Harvard University Press 1992); D. Stein, *International Diplomacy, State Administrators, and Narcotics Control* (Aldershot: Gower 1985); L.E.S. Eisenlohr, *International Narcotics Control* (London: George Allen and Unwin 1934); M. Greenberg, *British Trade and the Opening of China, 1800–1842* (Cambridge: Cambridge University Press 1951); W.O. Walker, *Opium and Foreign Policy: The Anglo-American Search for Order in Asia, 1912–1954* (Chapel Hill: University of North Carolina Press 1991); P.D. Lowes, *The Genesis of International Narcotics Control* (Genève: Librairie Droz 1966); J.Y. Wong, *Deadly Dreams: Opium, Imperialism, and the Arrow War (1856–1860) in China* (Cambridge: Cambridge University Press 2002).

2. Article 23c of the League Covenant; Article 295 of the Versailles Treaty (Germany); Article 247 of the St. Germain Treaty (Austria); Article 230 of the Trianon Treaty (Hungary); Article 174 of the Neuilly Treaty (Bulgaria); Article 280 of the Sevres Treaty (Turkey).
3. For a general overview, see W. McAllister, *Drug Diplomacy in the Twentieth Century. An International History* (London: Routledge 2000), Chapters 1–5.
4. The IO dealt with Burmese affairs until creation of the Burma Office in 1937.
5. The Dominions Division resided within the CO until elevated to the status of Dominions Office in 1925. The DD/DO served as an important conduit for relations with Australia, Canada, New Zealand, South Africa, and the DO also represented the South African High Commission Territories (Bechuanaland, Basutoland and Swaziland), Southern Rhodesia, the Irish Free State, Newfoundland, and (after 1918) Nauru.
6. For brief sketches of Campbell and Delevingne, see McAllister, *Drug Diplomacy*, pp. 48 and 51.
7. Original members of the League's Opium Advisory Committee included Great Britain, India, France, the Netherlands, China, Japan, Portugal, and Siam. Germany joined in 1922. India participated independently in important League-sponsored bodies such as the Mixed Sub-Committee on opium issues and an opium-related subcommittee of the League Health Committee. The Permanent Central Opium Board, created by the 1925 International Opium Convention, oversaw the licit trade and publicized instances of illicit trafficking: McAllister, *A Limited Enterprise*, pp. 54–8 and 125–30.
8. C. Terry and M. Pellens, *The Opium Problem* (New York: Bureau for Social Hygiene 1928), pp. 700–6; Gilchrist to Secretariat, 14 March 1922, League of Nations Archives (hereafter LNA), R710 12A/19019/10346; Delevingne to Drummond, 22 April 1921, LNA, R710 12A/12305/10346; Reports of Opium Advisory Committee sessions, League of Nations documents (hereafter LNds) A.38.1921.XI (3 September 1921); A.15.1922.[XI] (15 August 1922); A.15(a).1922.XI (2 September 1922); A.13.1923.XI (16 June 1923); proceedings of the League of Nations Health Committee and its opium subcommittee: LNds C.8.M.2.1922.III.; C.366.M.217.1922.III.; C.402.1922.III.; C.594.F.374.1922.III.; C.588.M.202.1924.II.

9. The 'Opium Confidential Print' covers the main themes of intra-imperial debate. See FO, *The Opium Trade, 1910–1941: Correspondence Respecting Opium*, PRO, 6 Volumes (Wilmington: Scholarly Resources 1974).
10. UK Public Record Office (hereafter PRO), CAB 24/168, C.P.450(24), 18 September 1924.
11. McAllister, *Drug Diplomacy*, pp. 57–67.
12. Interdepartmental conference records, 2 June 1924 (dated 21 June), PRO, FO 371/10324; multiple documents dated 21 July 1924–9 August 1924, PRO, FO 371/10325: F 2463/20/87; Delevingne to Waterlow, 2 August 1924, PRO, FO 371/10326; Collier and Newton minutes, 17 October 1924, PRO, FO 371/10327; Cabinet Papers, 24 July, 1924–11 October 1924, PRO, CAB 24/168.
13. PRO, FO 371/10327 (multiple documents dated 15–17 October 1924); FO minute, 21 October 1924, PRO, FO 371/10328.
14. McAllister, *A Limited Enterprise*, pp. 91–105.
15. Ibid., pp. 105–15.
16. FO, *The Opium Trade, 1910–1941*, 5, XXII, 1925, nos 10, 14, 18, and 6, XXIII, 1926, nos 5–9, 11; G. Blue, 'Opium for China: The British Connection' in Brook and Wakabayashi (eds), *Opium Regimes*, pp. 42–47.
17. Crosby to FO, 2 September 1927, PRO, FO 371/12533; Interdepartmental Opium Committeed (hereafter IOpC) meetings, 28 March 1928 and 19 February 1931, PRO, HO 45/20413; Grahame to FO, 9 November 1927, PRO, FO 371/12533; British Library of Information to FO, 4 May, 1 June, and 15 June 1928; Delevingne to Strang, 18 May 1928; Delevingne to Mounsey, 17 December 1928, all in PRO, FO 371/13260; R. Maule, 'The Opium Question in the Federated Shan States, 1931–36: British Policy Discussions and Scandal' *Journal of Southeast Asian Studies*, 23, 1992, pp. 14–36.
18. FO, The Opium Trade, 1910–1941, 6, XXIII, 1926, nos 1, 4, 17, 20, and 6, XXIV, 1927, nos. 6 and 8; PRO, FO 371/11715 (November–December 1926); PRO, FO 371/12526 (February–July 1927); PRO, HO 45/20413 (December 1926–June 1927); HO to FO, 29 June 1929 and Strang to Cadogan, 9 July 1929, PRO, FO 371/13976; CO to FO, 20 May 1930, PRO, FO 371/14772. According to circumstances, the IOpC met as frequently as several times a month, or as little as once a year.
19. Delevingne to Mounsey, 15 June 1926 and following documents in PRO, FO 371/11713:2479/604/87; Delevingne to Mounsey, 27 August 1926 and following documents in PRO, FO 371/11713:3532/604/87; Tilley to FO, 18 November 1926, PRO, FO 371/11714; J.M. Jennings, *The Opium Empire: Japanese Imperialism and Drug Trafficking in Asia, 1895–1945* (Westport, CT; London: Praeger 1997), pp. 73–5.
20. McAllister, *Drug Diplomacy*, pp. 95–102, 106, 123, and 127.
21. Ibid., pp. 114–118.
22. See, for example, seventeenth OAC session minutes, LNd C.661.M.136. 1933.XI (5 December 1933); Kinder and Walker, 'Stable Force in a Storm: Harry J. Anslinger and United States Narcotic Foreign Policy, 1930–1962', *Journal of American History*, 72, 4, 1986, pp. 917–8.
23. McAllister, *A Limited Enterprise*, pp. 184–96; Brook and Wakabayashi (eds), *Opium Regimes*; Walker, *Opium and Foreign Policy*; Jennings, *The Opium Empire*; E.R. Slack, *Opium, State, and Society: China's Narco-Economy and*

the Guomindang, 1924–1937 (Honolulu: University of Hawaii Press 2001); K. Meyer and T. Parssinen, *Webs of Smoke: Smugglers, Warlords, Spies, and the History of the International Drug Trade* (Lanham, MD: Rowman & Littlefield 1998).

24. See, for example, IOpC meetings for 1931–35 in HO 45/20413 and for 1935–41 in HO 45/20414.
25. IOpC meetings of 16 September 1930, 10 December 1930, 31 March 1931, 1 May 1933, 6 July 1933, 20 November 1933, all in PRO, HO 45/20413; IOpC meeting, 11 March 1935, PRO, HO 45/20414. The controversy involved Messrs. Bellaire, Atkinson, and Company, the only licensed, solvent British firm operating in Persia by the mid-1930s. IOpC records do not indicate the final disposition of this matter, but the case illustrates the type of non-medical factors that influenced imperial opium policymaking.
26. McAllister, *Drug Diplomacy*, pp. 128–33 and 148–9; McAllister, *The International Gaze*, mss.
27. McAllister, *Drug Diplomacy*, pp. 149–52.
28. IOpC meeting, 7 September 1943, PRO, HO 45/20414.
29. R. Cribb, 'Opium and the Indonesian Revolution' in *Modern Asian Studies* 22, 1988, pp. 701–22; D. Porch, *The French Secret Services: From the Dreyfus Affair to the Gulf War* (New York: Farrar, Straus and Giroux 1995), Chapters 13–14; A. McCoy, *The Politics of Heroin: CIA Complicity in the Global Drug Trade* (Chicago, IL: Lawrence Hill Books 2003), pp. 15–8, 23, 93–115, and 127–248; Walker, 'Opium and Foreign Policy', pp. 162–4 and 176–86.
30. London Embassy to State Department (hereafter SD), 18 August 1945, U.S. National Archives (hereafter NA), RG 59, 511.4A5/8-1845; IOpC meeting, 20 September 1945, PRO, HO 45/20414; Thornton to SD, 26 August 1948, PRO, FO 371/88825:US1811/79; Stone to SD, 28 November 1949, NA, RG 59, 800.114 Narcotics/11-2849; Hoare to FO, 15 April 1957, PRO, FO 371/129981; HO to Gauntlett, 18 July 1957, PRO, FO 371/129980; Stephens to Gauntlett, 30 April 1958, PRO, FO 371/137060.

13
The Trade-Off: Chinese Opium Traders and Antebellum Reform in the United States, 1815–1860

Elizabeth Kelly Gray

Introduction

During the Opium War of the 1840s, a writer for the Unitarian *Christian Examiner* echoed many Americans when he lambasted the British for introducing the drug into the 'great nation' of China. Opium's progress, he wrote in 1841, was 'more desolating than that of the sword, for it spares no condition, and more fatal than the plague, for it destroys the soul'.[1] The year before, 'L. M.', in writing for the publication, had praised the Perkins Institution and Massachusetts Asylum for the Blind where, he wrote, one could see 'benevolence working in its highest and surest paths'.[2] The *Christian Examiner* staff seemed unaware of the irony: the Perkins Institution was named after Thomas Handasyd Perkins. He was a merchant whose success in the Chinese opium trade had enabled him, upon retirement, to become its benefactor. In addition to support for the school for the blind, Perkins helped raise $100,000 as a trustee of the Massachusetts General Hospital for an Asylum for the Insane,[3] he supported the Boston Female Asylum for orphaned girls and he was a major donor to the Boston Athenaeum which he served for several years as president.[4]

Given the often intense criticism of the Opium War in America, Senator John C. Calhoun charged the British with waging war to 'force a poisonous drug down the throats of an entire nation';[5] it may appear incongruous that an American opium merchant could become a pillar of society. But before the conflict Americans had little sympathy for Asian opium addicts and regarded the addiction, when they thought of it at all, simply as part of Asian culture. More importantly, the war inspired outrage rather than soul-searching among Americans as few realized that their countrymen were involved in the opium trade. Fewer still traced the money into their own communities. Therefore, they did not link the growth of their railroads, hospitals, universities, orphanages and anti-slavery societies to the opium traffic. This chapter will examine American responses to opium trading and the First Opium War, before going on to further explore the irony outlined above; that while large numbers in the United States became instant critics of the commerce in the drug during Britain's conflict with China on the issue, many of them enjoyed the benefits of opium profits that had been invested in American institutions.

American opium trading

British merchants began importing opium from India to China in 1773, a move that shifted the balance of trade in their favor.[6] American merchants, who began trading with China in 1784,[7] followed suit. As early as 1804 the American firm Perkins & Co. was preparing to enter the Chinese opium trade[8] and 2 years later the Baltimore brig *Eutaw* brought the first recorded American opium ship to China.[9] The trade persisted until the Opium War, with brief interruptions due to the Embargo of 1807, the War of 1812 and unfavorable market conditions in 1819.[10] When American merchants appealed to Congress for protection during the Opium War, they acknowledged their part in the 'extensive opium trade' which had rapidly increased from 3210 chests imported in 1816–1817 to 23,670 chests by 1832–1833.[11]

Few Americans in the United States were aware that some of their countrymen were involved in the Chinese opium trade. British merchants were there in much larger numbers and therefore attracted most of the attention. Jacques Downs has observed that in 'almost all innovations which risked confrontation with the Chinese' in the years preceding the Opium War, 'British private merchants led and the Americans followed—cautiously'.[12] Staying in the shadows had its benefits. Although the absence of government involvement left residents at Canton unprotected during the conflict, America's informal commercial presence put its participants in an enviable situation: Great Britain would fight the war and sustain domestic and international criticism for it, while Americans could criticize both parties in the conflict and profit from the situation.[13] Illicit American trade thrived with limited criticism, and because the British won the war, Americans never had to decide how far they would go to protect the commerce.

Acknowledgements of American participation in the opium trade appeared before the war but received little attention. In 1827, Benjamin

Ellis noted in the *Journal of the Philadelphia College of Pharmacy* that Americans participated in the Chinese opium trade despite its prohibition 'by severe edicts of government'.[14] In 1837, Edmund Roberts reported that a few American ships were 'engaged in this illegal traffic' and that they brought the drug from Turkey.[15] That same year, a writer for the *Boston Medical and Surgical Journal* described the illegal trade and added that:

> Our own countrymen, some of them reputed of great respectability, are indeed deeply implicated in this crime of freighting a poison, which leads to certain moral infatuation, degradation and death.[16]

Early in 1839, another writer for the journal asked its readers 'is it not generally understood that some American merchants, of great capital, have accumulated their wealth by forcing immense quantities of gum opium into China, contrary to the known laws of the empire?'[17] E.W. Stoughton provided the answer in *Hunt's Merchants' Magazine* a year later, during the war: most American merchants in China, he wrote, were 'free from the slightest suspicion of having trafficked in the interdicted drug'.[18]

American domestic attitudes to opium

American outrage at the British during the Opium War suggests strong anti-imperial sentiment, but a broader perspective reveals that a passionate response was far from the norm. Before 1839, American writing about opium warned of the risks of overdosing and occasionally of the nascent domestic problem of addiction but most publications focused on medicinal uses and did not include warnings. Nicknamed 'God's own medicine' opium was one of the most commonly prescribed drugs in America.[19] Rather than being a cure in itself opium's value was thought to lie in its ability to allay pain. It could also control symptoms that physicians did not understand.[20] In 1810, Dr James Thacher suggested that opium be produced domestically due to its 'intrinsic value' and high price.[21] The main concern involved the risk of overdose. In 1803, University of Pennsylvania medical student Franklin Scott acknowledged that 'excessive doses' of opium could be fatal and that even small quantities could 'produce death in those unaccustomed to its use'.[22] In 1832, a writer for the *Boston Medical and Surgical Journal* insisted that there was 'little doubt' that people frequently 'pay with their lives' from taking overdoses of opium.[23] Many overdoses were accidental; others showed that opium could be the weapon of suicide.[24]

American physicians increasingly warned of opium addiction in the 1830s, largely because the nation's industrialization made a sober and efficient workforce more crucial.[25] But in 1824, Willard Phillips correctly stated in the *North American Review* that 'very few persons, if any, in this country, abandon themselves to the use of opium as a luxury'. Phillips was reviewing Thomas de Quincey's *Confessions of an English Opium-Eater*,[26] and he saw de Quincey's stark depiction of addiction 'more as an object of taste and literary curiosity' than as a cautionary tale, because he did not fear 'this species of intemperance' taking hold in America.[27] In 1830s, in *Essay on Temperance*, Edward Hitchcock concurred that there were 'few genuine opium eaters among us' and 3 years later a writer for the *Boston Medical and Surgical Journal* agreed that 'There are not many in this country addicted to the free and constant use of opium.'[28] David Courtwright has estimated that before 1842 no more than 0.72 Americans per thousand were opium addicts.[29]

Americans therefore had little first-hand knowledge of opium addiction and before the Opium War had little sympathy with Asian users. But their hearts went out to the handful of American addicts. The *Boston Medical and Surgical Journal* described a typical American addict in 1833 who took a tincture of opium called laudanum. She began taking the drug when a doctor prescribed it 'to quiet some slight degree of nervous irritation', then came to depend on it for 'many years', though 'she had *never*... been in the least degree *overexcited* by it'. Nobody, not even her husband, suspected her addiction. The woman wanted to stop taking the drug but attempts to decrease her dosage made her ill.[30] Like this woman, most American addicts began taking the drug under a doctor's care and continued the practice to avoid withdrawal.[31] American addicts tended to have 'a strong sense of the impropriety' of the practice, as the article's author noted,[32] and a canny addict could even hide his condition from his physician.[33] The woman's ability to remain productive, so long as she had her laudanum, was also typical. The writer theorized that the woman's situation was the 'most common' type of opium addiction in America as well as 'that which calls most loudly for the sympathy and aid of the humane physician'.[34]

There is little evidence, however, to suggest that Americans sympathized with Chinese opium addicts before the Opium War. References to Asian addiction before the conflict tended to portray opium use simply as part of Eastern cultures rather than as a pernicious aspect of the landscape. Americans saw few similarities between Asian and American addicts. The author who described the female addict defined 'opium eaters' only as those 'who took it originally as a medicine for some nervous affection, and continue it from necessity, rather than from choice'. He had 'no knowledge' of people who 'take opium for purposes of unnatural excitement and inebriation' but asserted that such people 'need less of our sympathy, and would excite us less to exertions in their behalf'.[35] By the time he wrote, the stereotype of the hallucinating Asian opium addict was engrained in American minds. In 1803, Franklin Scott had noted that people in 'eastern nations' were said to 'use Opium as we do tobacco'.[36] In 1829, Benjamin Ellis informed readers of the *Journal of the Philadelphia College of Pharmacy* that 'the consumption of drugs' in China was 'said to be immense, for they are almost as fond of them as they are of food'. He also stated that opium was more 'a luxury than a medicine' in China and that the illegal drug's popularity, which extended to government officials, 'annually drains the country of eight millions of dollars'.[37] In 1832, a writer for the *Boston Medical and Surgical Journal* looked to Asia to challenge the notion that opium use caused constipation: 'opium-eaters of the East', he pointed out, 'are not compelled to use cathartic medicine'.[38]

Asian addiction suggested shamelessness to the American reader who contrasted the public opium dens of the East with the furtive consumption of the local user. 'There is a particular part of Constantinople, called Theriaky Tcharchiffy, or Opium Bazaar', noted a writer for the *Boston Medical and Surgical Journal*, 'to which, at a particular hour in the evening, the lovers of the drug are habituated to find their way, for the purpose of indulging in this exhilaration'.[39] 'Exhilaration' was another source of contrast. While American addicts took opium to avoid withdrawal and to maintain their routines, Asian addicts smoked the drug to 'produce intoxication' and became completely unproductive. The 'images which flit before his diseased imagination, are exquisite, brilliant, heavenly', Edmund Roberts noted of the Asian addict, and while under the influence of the 'pernicious indulgence... all care was banished'.[40] After smoking opium the addict would sleep for several hours.[41]

The war would inspire sympathy for the Chinese and condemnation for the traders. But before the conflict, without the shadow of moral opprobrium over them, China merchants focused on their commerce with little criticism and easily justified their participation in the opium trade. Where justifications were concerned there was strength in numbers. According to John Upton Terrell, John Jacob Astor sold opium because 'inasmuch as "everyone else does it", he saw no reason why he should not profit from the illicit business'.[42] But the merchants' defensiveness is palpable in their allusions to colleagues who refused to participate in the trade. Some in China hated the drug. In 1827 the *Journal of the Philadelphia College of Pharmacy* published a letter written by a critic in Canton, describing the 'enormous quantities' of opium in China, its contraband status and the rapid expansion of the habit.[43] In 1839, E.C. Wines noted that 'Most Americans who trade to China are more or less engaged in the opium traffic' but that merchant Nathan Dunn 'was never interested to the amount of a dollar in that illicit commerce'.[44] To defend their practices, opium dealers characterized merchants who abstained from the trade not as law-abiding, but as 'squeamish', thereby suggesting that the abstainers' behavior was excessive.[45] They nicknamed Olyphant & Co. 'Zion's Corner' because its partners declined the trade, and in April 1829, merchant Benjamin Wilcocks was clearly rankled that John Whitall had refused, on moral grounds, to captain a ship that contained opium. The refusal 'lessened him in my opinion', Wilcocks noted in a letter to fellow merchant John Latimer. 'When a Captain stipulates for the particular articles which he will take on my ship, why let him go *you know where* for a cargo.'[46]

Although the trade was illegal, Chinese enforcement was lax until 1839, when Commissioner Tse-hsü Lin began confiscating shipments of the drug and sparked the First Opium War.[47] That conflict revealed transactions of which most Americans had been ignorant. Before hostilities observers could have surmised that the Chinese accepted the trade and that if they cared to protest the British would end it. But in 1839, Chinese efforts to stop the trade and Britain's bellicose response to ensure its continuance revealed the trade's sinister reality. While the Chinese wanted the trade to end, British greed appeared to exceed morality, and their overwhelming technological superiority ensured their victory.

Commentators throughout the United States condemned both the opium trade and the Opium War during the conflict. Americans in China, however, saw a more complex situation. Although opium importation was illegal, enforcement before 1839 was lax and some Chinese officials actively participated in the trade. When Edmund Roberts visited China just a couple of years before the war he noted that the opium trade had 'assumed the appearance of a regular branch of commerce', although he himself considered opium 'one of the most destructive narcotics which the world ever knew'. He recorded the acceptance of the trade by the Chinese despite its demoralizing effects on their society and economy. 'They say 'it is a Josh Pigeon' (meaning that God hath so decreed it) and they cannot prevent it', he observed. Although China had 'the most rigorous laws to prevent [opium's] importation and use', the complicity of Chinese officers ensured that sales would remain brisk. They warned smugglers, for example, of impending raids.[48] As will be shown below, merchants offered other justifications as well, such as theories that opium's dangers were exaggerated and that Chinese officials were more upset with their unfavorable balance of trade than with the physical effects on addicts.

American domestic attitudes to the First Opium War

Americans in the United States and in China differed dramatically in their attitudes toward the war and neither perspective took everything into account. A comparison of the two sides reveals anti-imperial inclinations and the strongly imperial reality. Americans at home saw a Manichean situation of British villainy and Chinese victimization. 'What cause of grievance the British originally had, does not so clearly appear', penned a writer for the *Catholic Telegraph* in the spring of 1840, 'they smuggled opium contrary to law, and the Chinese destroyed it'.[49] Americans in China, meanwhile, overemphasized mitigating factors on their behalf. The merchants insisted that the trade endured 'with the knowledge and consent of the chief local authorities' and that when the Chinese government truly wished 'to abolish the trade, they possess[ed] ample power ... to do it effectually'.[50]

While US merchants in China defended the war, America's business community at home was appalled. Western merchants in China were outraged at the Chinese crackdown on the trade in which authorities confiscated the drug without compensating merchants for it and confined Western merchants to their warehouses and offices. In an 1839 plea to the US government for protection, American residents of Canton insisted that Lin had made them 'prisoners' and threatened their lives while he confiscated the opium, a 'robbery' that lacked 'the slightest ground for justification'.[51] In his memoirs, opium merchant Robert Bennet Forbes dwelled more on China's 'great opium seizure' than on Britain's response to the confiscation.[52] Fellow merchant Warren Delano rejoiced in October 1840 at news 'that the Emperor had severely reprimanded *Linn*' for 'maladministration'.[53] Great Britain should 'knock a little reason into this bigotted people', he insisted, 'and teach them to treat strangers with common decency'.[54]

Back home, however, the war even repulsed Americans who sought increased trade with China. In *Hunt's Merchants' Magazine*, a commercial journal based in New York City, E.W. Stoughton feared that the conflict could cause the 'utter annihilation' of the United States' 'valuable

China trade'.[55] Stoughton and Francis Wharton also objected on moral grounds. Stoughton noted that the life of an addict was 'a dreadful punishment'. He criticized the East India Company for ignoring 'considerations of morality or religion' and Parliament for privileging economic concerns over 'morality, justice, and national honor'.[56] Initially optimistic that Westerners could bring the 'blessings' of 'Christianity and civilization' to the Chinese,[57] Wharton turned gloomy after the conflict. East India Company officials had 'seize[d]' on opium to 'save their commerce from extinction', even though addicts were 'hideous to behold, deprived of their teeth, their eyes sunk in their heads in a constant tremor'. There was 'no more glaring violation of the law of nations', he insisted, 'than the successful attempt... to cram down [China's] throat, by force, an article which she had deliberately refused to receive'.[58] Writers for Baltimore's *American Farmer* also objected. Although one had gleefully considered the possibilities if the Chinese became 'as fond of [tobacco] as they now are of the intoxicating, poisonous opium', the publication characterized the Opium War as having been 'conceived in iniquity and brought forth in sin'.[59]

Religious communities at home and abroad also had different opinions on the situation. Missionaries in China were bitter about their confinement early in the conflict. The Reverend Samuel Wells Williams criticized Lin for making no distinction 'between those who traded [in opium] and those who have not'[60] and the Reverend Elijah Bridgman insisted that the British had displayed 'Great moderation and magnanimity' in China.[61] Bridgman also saw the conflict as providential, proclaiming that 'the God of nations is about to open a highway for those who will preach the word'.[62] American Christians who were not involved in missionary work, meanwhile, sympathized more with the Chinese. A writer for the Baptist *Christian Watchman* characterized the Chinese crackdown as 'barbarous' but also felt that they had been 'more sinned against than sinning' in the conflict.[63] Writers for the *Christian Examiner* agreed that the Chinese would benefit from hearing 'the sound of the gospel'[64] but criticized the British for 'poisoning a whole people' with a 'pernicious' and forbidden drug 'for purposes of national aggrandizement'.[65] Others did not temper their opinions. While a writer for the Quaker journal the *Friend* was amazed that men of 'enlightened minds' would deal in opium, the editor of Cincinnati's *Catholic Telegraph* preceded comments by the Reverend David Abeel with a caveat: 'Mr. Abeel writes under the influence of some prejudice against the Chinese', noted the editor, 'which prevents him from rendering full justice to their efforts for the suppression of the deadly trade in opium'.[66]

A writer for the Congregationalist *Boston Recorder* was more succinct, exclaiming, 'When has a Christian and civilized nation been engaged in a more disgraceful enterprise!'[67]

Like other missionaries in China, medical missionaries saw the conflict not as disgraceful, but divine. Dr Daniel Macgowan and his colleagues noted the increased accessibility of the Chinese population after the conflict and proclaimed that 'God, in his wise providences, has effected mighty changes in this hitherto sealed country'.[68] Dr Peter Parker, who ran an Ophthalmic Hospital in Canton, regarded the war 'not so much as an opium or an English affair, as a great design of Providence to make the wickedness of man subserve his purposes of mercy towards China'.[69] Parker's hospital did not exist solely to cure ailing Chinese. Because the hospital attracted Chinese who shied away from religious instruction, some Americans saw it as a lure in opening China to Christianity and trade. In a published address the Medical Missionary Society in China explained that many Chinese were willing to 'forego their prejudices' and 'accept assistance, wherever they can find it'. The society admitted that the hospital was not 'the most direct and appropriate manner of introducing the gospel to a heathen people' but asserted that it was 'likely to open one avenue through which some of the blessings of Christianity may flow... while all other avenues are fast closed against it'.[70]

Contributors to medical journals praised Parker for curing ailments in China but financial support from British and American audiences tended to be conditional. During Parker's fund-raising tour in 1841 and 1842, a committee in Liverpool in the United Kingdom 'deemed it best to delay taking any steps till, at all events, *a partial opening of the China trade should be heard of*'. A New York society also informed Parker that it was not yet the right time to provide funds and one $50 contribution constituted Parker's entire take from Philadelphia. The striking exception was Boston, where Parker amassed *five thousand five hundred and fifty dollars*![71] Rather than signifying a striking generosity, Boston's support may suggest a different approach to currying favor with the Chinese. Two years earlier a writer for the *Boston Medical and Surgical Journal* had explained that part of the purpose of Parker's hospital was to raise the United States 'in the confidence and esteem of the Chinese, which will tend to put our commerce... upon a more desirable footing'.[72] While other cities chose the stick, Boston chose the carrot. It seemed a sensible approach as the Chinese did approve of the hospital. In July 1839, two of Commissioner Lin's deputies discussed the conflict with the doctor and Lin asked him for medical advice, including how to cure opium addiction.[73] In 1842, Parker was able to avoid being harmed during a Canton riot because someone told the crowd that Parker was 'the doctor—a good man'.[74]

During his tour Parker addressed Congress and insisted that his hospital had God's blessing,[75] but few American policymakers saw Britain's war as providential. In March 1840, Congressman Caleb Cushing called for a US representative to 'put the American trade with China on a just and stable footing' by contrasting Americans' 'proper respect' for Chinese law with the Britons' 'outrageous misconduct'. Congressman Francis Pickens of South Carolina assured Cushing that he had no intention of 'mak[ing] common cause with England in her designs in China' or of helping to 'forc[e] on the Chinese the odious traffic in opium'.[76] When Congressman John Quincy Adams blamed the conflict on Chinese arrogance rather than on opium, he was roundly criticized.[77] In an 1841 lecture to the Massachusetts Historical Society, the former president characterized commerce as being among the 'natural rights and duties of men' and blamed the war on China's 'arrogant and insupportable pretension' to allow trade only 'upon the insulting and degrading forms of the relation between lord and vassal'.[78] John Gorham Palfrey had already accepted Adams's paper for publication in the *North American Review*, but he withdrew his offer after he heard the presentation.[79] Meanwhile, a writer for the *Christian Examiner* spent the better part of a 40-page article refuting Adams's assertions. Insisting that 'We have no right to force our commerce upon a nation... any more than they have to impose conditions on us' the writer was not appeased by predictions that the war could result in a Christianised China. 'I have no faith', he stated, 'in the promotion of morality by immoral means.'[80]

Because Americans in China and in the United States had such different perspectives on the Opium War, it is perhaps unsurprising that there was friction when the two groups met. One who doubtlessly experienced such heated encounters was Robert Bennet Forbes, a merchant who candidly described his involvement in the opium trade in his memoirs. One can glean from his writings that frequent criticism was the price for his candor. To Forbes, the opium trade was second nature; he had grown up in a merchant community, where the opium trade rarely received criticism. When he was 12-years old, he was doing odd jobs at a Boston trading house. At 13, he made his first trip to Canton, where his relatives were involved in the trade.[81] By 36, he had become 'head of the largest American house in China'.[82]

Two years after the war, Forbes defended the merchants' role to a society newly outraged at the traffic. In the 1844 *Remarks on China and the China Trade*, he noted that he was often asked about the opium trade and about what Great Britain gained ' "by her cruel and oppressive war" '.[83] Forbes acknowledged the war's 'bloody scenes' but defended the British by implying that foreign trade was the obligation of every nation. He presented the story of Captain Weddel, who centuries earlier had resorted to violence after failing to open China peaceably, as 'valid evidence that the only proper negotiators for the Chinese were iron balls and fire and sword'.[84] In a review of the book a writer for the *North American Review* chided Forbes for dealing with the opium trade 'only as a great commercial and political question; its moral aspect', he continued, 'in relation both to the consumers and the parties who bring it to China, being left entirely out of view'.[85]

In Forbes's *Personal Reminiscences*, which was published in 1876, the merchant's frustration was palpable as he provided a battery of excuses for the trade, before finally giving up. He noted that the opium trade could not have occurred without 'the connivance of the Chinese local authorities'. He naturalized the trade, stating that there was a 'regular understanding' between the merchants and the Chinese officials, and that the trade went 'harmoniously... like any honest traders', unless someone tried to ship opium 'without paying the mandarins'.[86] The opium trade was *technically* illegal but 'the officers of government encouraged its cultivation' for bribes and because many were addicts themselves. It was the unfavorable balance of trade, he continued, not the 'moral effect on the people', that caused the Chinese government to intervene. He then added that the trade imbalance was not as great as had been suggested and that opium's 'demoralizing' effects were 'probably' no greater than those caused by 'the use of ardent spirits'. None of the people involved regarded the trade as 'smuggling', he continued, explaining that 'it was viewed as a legitimate business so long as the drug was sold on the coast'; it was 'certainly legitimate in India... at Singapore, at Manila, [and] at Macao'. These are not the words of a man whose community had accepted his former line of work. At this point one can almost hear his exhaustion; 'I shall not go into any argument', he concluded in the section, 'to prove that I considered it right to follow the example of England, the East India Company... and the merchants to whom I had always been accustomed to look up as exponents of all that was honorable in trade'.[87]

Forbes encouraged colleagues to publish their recollections. In *'Fae Kwan' at Canton before Treaty Days*, published in 1882, W.C. Hunter played down opium's ill effects. He insisted that the Chinese were too productive, and too poor, to be addicts. They were 'healthy, active, hard-working, and industrious... intelligent in business [and] skilful in manufactures and handicrafts', traits that he argued were 'inconsistent with habitual smoking'. Meanwhile, he stated that the majority could only afford 'a dilution' of the drug, which was so weak as to be 'utterly harmless'. He admitted that usage was 'more or less common' among the wealthy but insisted that 'No evidences of a general abuse, rarely of the use of the pipe, were apparent.' He also echoed the notion that opium's evil consequences were 'infinitesimal' when compared with alcohol consumption in the West. In the epilogue, Hunter acknowledged that 'Robert B. Forbes, Esq., of Boston, U.S.A.' had provided 'untiring encouragement and assistance' to him to recall 'the days of Old Canton'.[88]

Merchants and philanthropy in America

Despite the efforts of those like Forbes and Hunter, the American public remained largely critical of the war and of the opium trade when the subject came up. Opium merchants who retired before the Opium War, and who were comparatively quiet about their business dealings, avoided the criticism that Forbes encountered. Despite their willingness to become involved in the opium trade, they were also well-respected members of society. Stephen Girard, for example, declared in 1806 that he was 'very much in favor of investing heavily in opium'.[89] By 1820, he was shipping as much as 7500 pounds of the drug per voyage and controlled almost half of the trade passing through Batavia (now Jakarta) in Indonesia.[90] At a time when successful merchants in China could earn about $100,000 in 7 years, and thereby go home wealthy,[91] Girard amassed more than $7 million before his death in 1831. For the last 20 years of his life, he was America's wealthiest man.[92] His fortune enabled him to rescue the US government financially during the War of 1812 and to help create the Second Bank of the United States.[93] Preceded in death by his wife, Girard also died childless. He bequeathed his millions for public use; his will was the first American document that outlined philanthropy on a large scale.[94] Girard's legacy to the Pennsylvania Hospital was the largest that the institution received in its first hundred years and one of only two contributions that far exceeded five thousand dollars.[95] He gave hundreds of thousands of dollars to improve the state's canals and to pave and light streets, and tear down wooden buildings, in the City of Brotherly Love.[96] Most notably, he allotted $2 million to endow Girard College, a school for orphans in Philadelphia.[97] Girard College became the largest boarding school in the world.[98]

The aforementioned Thomas Handasyd Perkins was another opium merchant who retired before the Opium War and became a respected philanthropist, and the Perkins Institution for the Blind is a prime example of the opium trade's contributions to American society. Dr Samuel G. Howe tutored six blind students at the New England Institute for the Blind in the early 1830s and demonstrated 'their capacity for improvement' to the public. At this time, according to Howe, the school's 'Treasury was empty, and the Institution in debt' by hundreds of dollars.[99] Perkins, by then retired from the trade, was 'deeply interested' by the display. He donated his Pearl Street mansion to the institute and spurred a community drive to raise $50,000 for the school.

The re-christened Perkins Institution and Massachusetts Asylum for the Blind was the only school for the blind at that time in New England and became one of the best-known charitable institutions in the state.[100] Its most celebrated student was the deaf, blind and mute Laura Bridgman, whose accomplishments disproved the conventional wisdom of the time, that such a person could not receive 'effectual instruction in reading and writing'.[101] Howe communicated with Bridgman by teaching her a finger alphabet. Charles Darwin wrote about Bridgman, and Charles Dickens devoted a long section of his *American Notes* to her case.[102] When Helen Keller's mother read the Bridgman chapter in *American Notes*, she contacted the Perkins school to get a teacher for her daughter. Bridgman had taught the finger alphabet to Anne Sullivan, a student at the school who became Keller's miracle worker.[103] Perhaps most important, the Perkins Institution's renown encouraged more widespread education of blind children from non-elite families.[104]

Americans tended not to connect the philanthropy with its illicit source, likely because they did not even know that the benefactors were involved in the opium trade. Their ignorance made the opium issue appear simpler than it actually was, because they did not have to weigh opium's negative effects in China with the benefits of opium fortunes in America. As has been noted, some observers even condemned the opium trade while praising the institutions that the trade helped to sustain. In March 1838, a writer for the *Boston Medical and Surgical Journal* described the 'habitual opium-eater' as having 'total attenuation of body, a withered, yellow countenance, a lame gait, [and] a bending of the spine' and added that 'These people seldom attain the age of forty if they have begun to use opium at an early age'. That same month a writer for the periodical praised the New England Institution for the Education of the Blind for opening 'avenues of happiness' to its students and insisted that the school's most recent report 'cannot be read without exciting the liveliest emotions in every benevolent breast'.[105] Just as in the *Christian Examiner*, there was seemingly no awareness that the institution for the blind owed its prosperity to an opium fortune. Perhaps this was because Perkins retired long before the Opium War or because, unlike Forbes, he never published candid memoirs of his career. The only contemporary biography of Perkins, a lengthy article in *Hunt's Merchants' Magazine* that was published in an expanded form as the *Memoir of Thomas Handasyd Perkins* in 1856, was written by his son-in-law and never mentioned opium.[106]

But the source of Perkins's fortune was illicit. 'It is our intention', Perkins wrote to a colleague of the contraband trade in 1818, 'to push it as far as we can'.[107] A Perkins & Co. letter from August 7, 1819 acknowledged 'the stigma that attaches to those who deal' in the drug. 'It is considered a very *disreputable* business' the author explained '& view'd by the Chinese' as akin to 'smuggling'.[108] An 1824 letter acknowledged difficulties with the 'illicit trade' due to 'existing prohibitions in China' and advised the crew of a company vessel 'to judge how far this trade can be prosecuted with safety'.[109] By 1830, Perkins had personal wealth valued at more than $700,000 acquired, as recent biographers have put it, because 'China, and opium, had been good to him'.[110] Perhaps Perkins himself best expressed the drug's importance in an 1827 letter: 'I have written and thought so much of Opium' he wrote 'that it gives me an opiate to enter upon the subject'.[111]

Indeed, even in the wake of the Opium War that Americans had condemned so vigorously, Sino-American trade grew and American opium merchants kept giving money away. Robert Bennet Forbes did humanitarian work such as 'establishing coastal life-saving units and homes for aged sailors'. He also brought food to the Irish during the potato famine.[112] Opium philanthropists in the antislavery cause included Forbes's younger brother, John Murray Forbes. During the battle for Kansas in the 1850s the younger Forbes organized the New England Emigrant Aid Society, which supported antislavery settlement in the new state, and he helped fund John Brown's violent campaign in Kansas.[113] He also supported experiments that prepared for free labor on plantations and chaired the Committee of One Hundred which promoted the use of African American troops to fight for the Union.[114] When John C. Green became head of Russell & Co. in January 1834, his 'rigorous reorganization' of the company probably saved it. He went on to become 'the most generous philanthropist in the history of Princeton University up to the end of the 19th century'.[115] Other institutions that benefited from opium merchants' largesse included Massachusetts

General Hospital, the Boston Athenaeum, the Butler Hospital for the Insane, Harvard University and the Antislavery Society. Also, retired opium merchants served as presidents of the Society of the Cincinnati and the Boston Athenaeum.

Opium money also hastened the progress of American industry. Thomas H. Perkins sponsored construction of what was arguably the first railroad in America.[116] Stephen Girard invested in the construction of bridges and railroads, including the precursor to the Reading Railroad.[117] In 3 years, John Murray Forbes raised more than $6 million for the Michigan Central Railroad, progress on which had been stalled by the Panic of 1837. He was also instrumental in the development of the Chicago, Burlington & Quincy Railroad, which spanned four states with almost 3000 miles of track.[118] In so doing, investors like Forbes advanced industrial development that otherwise may have languished. In 1859, Iowa Governor Ralph Lowe asked his constituents, many of whom resisted outside control from investors such as Forbes, 'How are our railroad schemes to be carried forward' in the absence of such investments?[119]

Such significant economic contributions lead one to what Jacques Downs has called 'the riddle of the virtuous American opium trader in China' in which each merchant he studied had 'an impeccable business and moral reputation at home'.[120] The very involvement in a trade considered so murky may have made necessary the philanthropic gestures as the merchants sought respectability. For merchants in the early nineteenth century there was a growing stigma even *without* opium; Boston's elite perceived merchants as having 'closed hearts, and tight fists'. As the elite began opting for professions that were 'untainted with any suggestion of money-making', Perkins's 'compulsive, entrepreneurial spirit' became precisely the quality that those who did pursue business hoped to avoid.[121] This antagonism is shown in the stories of T.S. Arthur, who wrote for *Godey's Lady's Book*, the most popular women's magazine of the day. In 'Marrying a Merchant' Josephine plans to marry a merchant because she wants 'to be well taken care of, and to be sustained in good society'. She recommends that Mary do the same, and specifically suggests a merchant named Perkins, but Mary instead marries a clerk whom she loves. Josephine laughs at her friend's choice, but Mary ends up happily married and sufficiently well off, while Josephine's husband only married her to have a 'showy' wife, and the couple lives beyond their means.[122] To distance themselves from this greedy image, many merchants including, most prominently, Thomas H. Perkins developed strong amateur interests in horticulture to demonstrate their ability 'to appreciate what had no market value and no apparent uses'.[123]

That is not to say that the merchants thought of themselves as immoral. Jonathan Goldstein has argued that one must look at 'the mores of the age' to understand how a man like John Latimer 'could have engaged in the opium trade' when he was 'an otherwise upright individual' who was president of the Society of the Cincinnati, cared for a girl when her merchant father abandoned her and was 'an outspoken advocate of mercantile ethics in other areas'.[124] This was a period when the slave trade, the coolie trade and the removal of Indians were just some of the activities that resulted in 'direct, visible human suffering for which the entrepreneur was clearly responsible'.[125] Traders also had their own code of ethics. They had traveled halfway around the world with other men's money and their 'primary duty' was 'to preserve, and if possible, to increase the property with which they had been entrusted'.[126] Indeed, opium merchants probably grew up reading sea stories in which merchant-adventurers were 'exempt from most moral rules' because trade was 'a contest in which one either wins or loses'.[127] Trading with the Chinese would have been 'outside the bounds of the moral code' because they were not Christian, they appeared corrupt and they treated Americans with contempt. Just as with the slave and coolie trades and the removal of Indians, those in charge could not identify with those they affected; 'once the Chinese were defined as alien and somehow bad ... the opium trade became psychologically possible'.[128] For this reason, Dr Daniel Macgowan's warning was likely spurious, when he told a New York physicians' temperance society in 1840 that, if Chinese officials banned opium from their nation, 'the followers of the prophet would find a market for their staple poison on our own shores'.[129] The merchants were willing to treat Chinese society in a way that they would never have contemplated approaching their own. Some, like Robert Bennet Forbes, tried to justify this through argument with critics; most preferred instead to camouflage their positions by philanthropic largesse.

Conclusion

This chapter has demonstrated that American opinion was far from united toward opium trading in the first half of the nineteenth century. Those whose fortunes, either as traders or missionaries, were tied to the success of the British conflict with China over opium were considerably less critical of opium trading than those who had little direct interest in its transactions. Indeed, there was little interest in the commerce in the drug before the First Opium War and most remained ignorant of the extent to which Americans were involved in trading in what was often considered at home to be 'God's own medicine'. When Britain and China went to war on the issue Americans in the United States tended to criticize the British, among the most persistent critics were American merchants who saw opportunity in the difficulties of a trading rival in the Chinese market and the religious who readily accepted that China's case was a moral one.

However, the general ignorance of opium trading on the part of those engaged in criticizing it meant that one crucial aspect of the commerce that directly affected many of those in the United States went unnoticed. Attention to merchants' philanthropy is not intended to suggest that they were selfless. They were not. They were in business to make money and they kept most of what they made. But such attention reveals an important aspect of the opium trade on which contemporary sources are largely silent: the fact that the huge wealth accumulated by American opium traders had a significant impact on communities in the United States. This raises the interesting question of whether Americans would have been less critical of the opium trade if they had balanced its perceived drawbacks with its benefits for the mentally ill, the fillip given to the nation's universities, or its help to the blind in 'offer[ing] intelligence, enjoyment, and usefulness, in place of ignorance, sorrow, and idleness'?[130]

In February 1839, Dr J.V.C. Smith, perceiving an increase in drug addiction in America, warned readers of the *Boston Medical and Surgical Journal* that 'the secret consumers of opium in the United States are vastly more numerous than is suspected'.[131] There were actually even more 'secret consumers' than Smith himself realised. Although he was referring to habitual users of laudanum, he could have included patrons of the Boston Athenaeum, patients at Massachusetts General Hospital, students and professors at Princeton and Harvard, and orphans at Girard College. As the era of social reform in the United States developed, more Americans were becoming addicted to the opium profits of empires abroad than they could have ever imagined.

Notes

1. 'The Opium War, and its Justice' in *Christian Examiner*, May 1841, p. 224.
2. L.M., 'The Perkins Institution', in *Christian Examiner*, July 1840, p. 378.
3. T.G. Cary, *The Memoir of Thomas Handasyd Perkins* (Boston: Little, Brown and Company, 1856), p. 220.

4. D. Gold Hansen, *Strained Sisterhood: Gender and Class in the Boston Female Anti-Slavery Society* (Amherst: University of Massachusetts Press 1993), pp. 58–60.
5. J.C. Calhoun, *The Papers of John C. Calhoun*, volume 15 (Columbia, SC: University of South Carolina Press 1959–99), p. 154; 'Congress' in *Emancipator*, March 19, 1840, p. 187.
6. C. Seaburg and S. Paterson, *Merchant Prince of Boston: Colonel T.H. Perkins, 1764–1854* (Cambridge, MA, 1971), p. 264; R.B. Forbes, *Remarks on China and the China Trade* (Boston: Samuel N. Dickinson 1844), p. 45.
7. T.N. Layton, *The Voyage of the 'Frolic': New England Merchants and the Opium Trade* (Stanford, CA: Stanford University Press 1997), pp. 26–30.
8. J. Goldstein, *Philadelphia and the China Trade, 1682–1846: Commercial, Cultural, and Attitudinal Effects* (University Park, PA: The Pennsylvania State University Press 1978), p. 52.
9. Ibid., p. 53.
10. C.C. Stelle, 'American Trade in Opium to China, Prior to 1820' in *Pacific Historical Review*, December 1940, pp. 434–5 and 444.
11. 26th Cong., 1st sess., Doc. No. 40.
12. J.M. Downs, 'American Merchants and the Chinese Opium Trade, 1800–1840' in *Business History Review*, 42, 1968, p. 429.
13. Although it is beyond the scope of this article, many Britons criticized their government's prosecution of the Opium War. For Forbes, America's best-of-both-worlds situation was not enough; he was piqued that the United States was excluded from discussions regarding which ports to open to foreign commerce (Robert B. Forbes, *Personal Reminiscences* (Boston 1882), p. 364).
14. B.E., 'Opium' in *Journal of the Philadelphia College of Pharmacy*, November 1827, pp. 110–1.
15. E. Roberts, *Embassy to the Eastern Courts of Cochin-China, Siam, and Muscat; in the U. S. Sloop-of-War Peacock* (New York: Harper & Brothers 1837), pp. 66 and 126.
16. 'Opium Eating in Siam' in *Boston Medical and Surgical Journal*, April 26, 1837, pp. 193–4.
17. 'Premium Essay on the Opium Trade' in *Boston Medical and Surgical Journal*, February 13, 1839, pp. 18–19.
18. E.W. Stoughton, 'The Opium Trade—England and China' in *Hunt's Merchants' Magazine*, May 1840, p. 386. Stoughton's reference to Americans being 'free from the slightest suspicion' of importing opium may be his clever way of implying that Americans were innocent, but he may not have known they were guilty. Writers for *Hunt's* acknowledged American participation in the trade that December by using a document that only became available after Stoughton's article was published (Exec. Doc. 26/1, #248).
19. D. Courtwright, *Dark Paradise: A History of Opiate Addiction in America* (Cambridge, MA: Harvard University Press 2001), p. 45; H. Wayne Morgan, *Drugs in America: A Social History, 1800–1980* (Syracuse, NY: Syracuse University Press 1981), p. 1.
20. Morgan, *Drugs in America*, p. 3.
21. J. Thacher, *The New American Dispensatory* (Boston: T.B. Wait 1810), pp. 454–7, quoted in Morgan, *Drugs in America*, p. 2.

22. F. Scott, *Experiments and Observations on the Means of Counteracting the Deleterious Effects of Opium, and on the Method of Cure of The Disease Resulting Therefrome* (Philadelphia: Maxwell 1803), p. 7.
23. 'Effects of Opium Eating' in *Boston Medical and Surgical Journal*, April 4, 1832, pp. 130–1.
24. Morgan, *Drugs in America*, p. 6.
25. Courtwright, *Dark Paradise*, p. 45; Morgan, *Drugs in America*, pp. 8–9.
26. J. North, *De Quincey Reviewed: Thomas de Quincey's Critical Reception, 1821–1994* (Columbia, SC: Camden House 1997), p. 11.
27. W. Phillips, 'Confessions of an English Opium-Eater' in *North American Review*, January 1824, p. 92.
28. E. Hitchcock, *An Essay on Temperance, Addressed Particularly to Students, and the Young Men of America* (J.S. & C. Adams 1830), p. 8; 'Opium Eating' in *Boston Medical and Surgical Journal*, September 4, 1833, p. 66.
29. Courtwright, *Dark Paradise*, p. 28.
30. 'Opium Eating', pp. 66–7.
31. Courtwright, *Dark Paradise*, p. 42.
32. 'Opium Eating', p. 66.
33. 'Effects of Opium Eating' in *Boston Medical and Surgical Journal*, April 4, 1832, pp. 128–9.
34. 'Opium Eating', p. 67.
35. Ibid., p. 66.
36. Scott, *Experiments and Observations*, p. 20.
37. B.E., 'Chinese Materia Medica' in *Journal of the Philadelphia College of Pharmacy*, July 1829, pp. 151–52.
38. 'Effects of Opium Eating', p. 131.
39. 'Opium', *Boston Medical and Surgical Journal*, March 22, 1837, p. 101.
40. Roberts, *Embassy to the Eastern Courts*, pp. 139, 151 and 142.
41. G.H. Smith, 'Opium Smoking in China' in *Boston Medical and Surgical Journal*, May 25, 1842, pp. 246–7. The *American Journal of the Medical Sciences* also carried Smith's account ('Abstract of a paper on Opium smoking in China', July 1842, pp. 229–33).
42. J. Upton Terrell, *Furs by Astor* (New York: William Morrow & Company 1963), p. 292.
43. Quoted in B.E., 'Opium' in *Journal of the Philadelphia. College of Pharmacy*, November 1827, pp. 110–1.
44. E.C. Wines, *A Peep at China, in Mr. Dunn's Chinese Collection* (Philadelphia: Printed for Nathan Dunn 1839), p. 10.
45. Terrell, *Furs by Astor*, p. 292.
46. Quoted in Goldstein, *Philadelphia and the China Trade*, p. 50 (Original in Latimer Papers, Library of Congress, April 26, 1829).
47. C. Trocki, *Opium, Empire and the Global Political Economy: A study of the Asian opium trade, 1750–1950* (London: Routledge 1999), pp. 97–100; H.W. Brands, *The United States in the World: A History of American Foreign Policy*, volume 1 (Boston: Houghton Mifflin Company 1994), pp. 119–21.
48. Roberts, *Embassy to the Eastern Courts*, pp. 139–41.
49. Late and Important from China' in *Catholic Telegraph*, March 28, 1840, p. 103.
50. 26th Cong., 1st sess., Doc. No. 40.

51. Ibid.
52. Forbes, *Personal Reminiscences*, p. 346.
53. WD, Macao, to RBF, October 11, 1840. WDP, Franklin Delano Roosevelt Library, Hyde Park, NY.
54. WD, Canton, to Frederic H. Delano, New York City, September 21, 1839, WDP, FDR Library.
55. Stoughton, 'The Opium Trade', p. 386.
56. Ibid., pp. 387–8, 391 and 396.
57. F. Wharton, 'East India and the Opium Trade' in *Hunt's Merchants Magazine*, January 1841, pp. 17, 21–2
58. Francis Wharton, 'China and the Chinese Peace' in *Hunt's Merchants Magazine*, March 1843, pp. 205–6, 222–4 and 226.
59. 'Substitute for Opium' in *American Farmer*, May 20, 1840, p. 412; 'Latest from Europe' in *American Farmer*, October 21, 1840, p. 175.
60. 'China:—Letter from Mr. Williams' in *Missionary Herald at Home*, December 1839, p. 464.
61. 'China' in *Missionary Herald at Home*, November 1841, pp. 471–72.
62. Quoted in 'Other Societies:—American Board of Commissioners for Foreign Missions: China' in *Baptist Missionary Magazine*, February 1841, p. 52.
63. 'War in China' in *Christian Watchman*, January 15, 1841, p. 9.
64. 'Malcom's Travels' in *Christian Examiner*, July 1839, p. 387.
65. 'The Opium War and its Justice', pp. 224–9.
66. 'The Opium Trade in China' in *Friend*, February 29, 1840; 'Foreign: New by the Taralinta: China' in *Catholic Telegraph*, March 7, 1840, p. 75.
67. 'England and China' in *Boston Recorder*, July 10, 1840, p. 110.
68. 'American Baptist Board of Foreign Missions' in *Baptist Missionary Magazine*, December 1843, p. 316.
69. 'Recent Intelligence' in *Missionary Herald at Home*, January 1841, p. 43; Parker's quote also appeared in 'Other Societies: American Board of Commissioners for Foreign Missions: China' in *Baptist Missionary Magazine*, February 1841, p. 52.
70. 'Medical Missionary Society in China' in *Missionary Herald at Home*, March 1839, p. 113.
71. 'Medical Missionary Hospital in China' in *Boston Medical and Surgical Journal*, January 24, 1844, p. 506.
72. 'Medical Missionary Society in China' in *Boston Medical and Surgical Journal*, January 30, 1839, p. 416.
73. 'Dr. Parker' in *Boston Medical and Surgical Journal*, February 19, 1840, p. 34.
74. Quoted in 'China: Letter from Doct. Parker' in *Missionary Herald at Home*, June 1843, p. 257. Mrs Parker's account of the incident appears in 'Interesting Letter from China' in *Boston Recorder*, April 27, 1843, p. 68.
75. J. Spence, *To Change China: Western Advisers in China, 1620–1960* (New York: Penguin Books 1980), p. 49.
76. *Cong. Globe*, 26th Cong., 1st sess. (March 16, 1840), pp. 275.
77. *Cong. Globe*, 26th Cong., 2d sess., pp. 28–29.
78. Adams quoted in Walter LaFeber (ed.), *John Quincy Adams and American Continental Empire: Letters, Papers, and Speeches* (Chicago: Quadrangle Books 1965), pp. 50–1.

79. G.C. Ward and F.D. Grant, 'A Fair, Honorable, and Legitimate Trade' in *American Heritage*, August–September 1986, p. 57; Paul C. Nagel is reaching when he portrays Adams's speech as a success (Nagel, *John Quincy Adams: A Public Life, A Private Life* (New York: Alfred A. Knopf 1997), pp. 383–4); Jacques M. Downs, 'Fair Game: Exploitive Role-Myths and the American Opium Trade' in *Pacific Historical Review*, 41, 1972, pp. 144–5.
80. W.A., 'Great Britain and China' in *Christian Examiner*, July 1842, pp. 298 and 318.
81. Forbes, *Personal Reminiscences*, pp. 27–28, 31 and 41.
82. Ibid., pp. 168–9.
83. Forbes, *Remarks*, p. 4.
84. Ibid., p. 36.
85. 'Forbes on China and the China Trade' in *North American Review*, October 1844, p. 494; Forbes asserted that he wanted 'more able hands' to address the trade's moral aspects, confined his memoir to 'facts', and allowed readers 'to form their own conclusions' (Forbes, *Remarks*, p. 44).
86. Forbes, *Personal Reminiscences*, p. 143.
87. Ibid., pp. 143–5.
88. W.C. Hunter, *The 'Fae Kwan' at Canton before Treaty Days, 1825–1844* (Taipei: Ch'eng-wen Publishing Company 1965 (originally published 1882), pp. 79–80 and 158.
89. Stephen Girard to Mahlon Hutchinson and Myles McLeveen, January 2, 1806, quoted in Goldstein, *Philadelphia and the China Trade*, pp. 47 and 54.
90. H.E. Wildes, *Lonely Midas: The Story of Stephen Girard* (New York: Farrar & Rinehart 1943), p. 171.
91. J.M. Downs, 'The Commercial Origins of American China Policy, 1784–1844' in Jonathan Goldstein, Jerry Israel and Hilary Conroy (eds), *America Views China: American Images of China Then and Now* (Bethlehem: Lehigh University Press 1991), p. 58.
92. J.G. Lee, *Philadelphians and the China Trade, 1784–1844* (Philadelphia: University of Pennsylvania Press 1984), p. 104.
93. Goldstein, *Philadelphia and the China Trade*, p. 42; Lee, *Philadelphians and the China Trade*, 104.
94. Lee, *Philadelphians and the China Trade*, p. 104; Goldstein, *Philadelphia and the China Trade*, p. 44.
95. G.B. Wood, M. D., *An Address on the Occasion of the Centennial Celebration of the Founding of the Pennsylvania Hospital* (Philadelphia: T.K. and P.G. Collins 1851), p. 126.
96. C.A. Herrick, *Stephen Girard: Founder* (Philadelphia: Girard College 1923), pp. 188–9.
97. Ibid., pp. 179 and 184.
98. Wildes, *Lonely Midas*, p. 315.
99. S.G. Howe, 'Education of the Blind' in *North American Review*, July 1833, p. 57; M. Howe and F. Howe Hall, *Laura Bridgman: Dr. Howe's Famous Pupil and What He Taught Her* (Boston: Little, Brown, and Company 1903), p. 10.
100. Cary, *The Memoir of Thomas Handasyd Perkins*, p. 223; *The National Almanac and Annual Record for the Year 1863* (Philadelphia: George W. Childs 1863), p. 374 (Making of America database, University of Michigan, hereafter MOA); L. Colange, *Zell's Popular Encyclopedia, A Universal Dictionary*

of English Language, Science, Literature, and Art, volume 2 (Philadelphia: T. Ellwood Zell, 1871), p. 312 (MOA).

101. Cary, *The Memoir of Thomas Handasyd Perkins*, p. 224.
102. Darwin wrote about her in *The Expressions of the Emotions in Man and Animals* (Elisabeth Gitter, *The Imprisoned Guest: Samuel Howe and Laura Bridgman, the Original Deaf-Blind Girl* (New York: Farrar, Straus and Giroux 2001), p. 5). Dickens describes the Perkins Institution, Laura Bridgman, and her fellow student Oliver Caswell in *American Notes* (Gloucester, MA: Peter Smith 1968), pp. 44–61.
103. Gitter, *Imprisoned Guest*, pp. 6–8.
104. I. Mayhew, *Popular Education: For the Use of Parents and Teachers, and for Young Persons of Both Sexes* (New York: Harper & Brothers 1850), p. 124 (MOA).
105. 'Opium-Eating' in *Boston Medical and Surgical Journal*, March 28, 1838, p. 129; 'Education of the Blind' in *Boston Medical and Surgical Journal*, March 21, 1838, p. 101.
106. Seaburg and Paterson, *Merchant Prince of Boston*, p. 420. References to Perkins in secondary literature rarely mention opium, partially because sources tend to use the Cary *Memoir*. In its entry for Perkins, the 1946 *Dictionary of American Biography* mentioned the China trade but described Perkins only as a 'merchant philanthropist' and as someone who was 'best known for his philanthropies' (D. Malone (ed.), *Dictionary of American Biography*, volume 14 (New York: C. Scribner's Sons 1946), p. 477). Perkins was referred to as 'an eminent merchant and a distinguished citizen' *The Memorial History of Boston, 1630–1880*, volume 4 (Boston: Ticknor and co., 1880), p. 207. Recent biographies of Laura Bridgman refer to Perkins only as 'a shipping tycoon' and 'a wealthy merchant' (Gitter, *Imprisoned Guest*, p. 43; E. Freeberg, *The Education of Laura Bridgman: First Deaf and Blind Person to Learn Language* (Cambridge, MA: Harvard University Press 2001), p. 17).
107. 'Extracts', March 24, 1818, quoted in Seaburg and Paterson, *Merchant Prince of Boston*, p. 285.
108. Perkins & Co. letter, Canton, August 7, 1819, Perkins Papers, Massachusetts Historical Society (MHS).
109. J.L. Perkins and T.H. Perkins, Boston, to Perkins & Co., Duxbury, April 28, 1824, Perkins Letters, MHS.
110. Seaburg and Paterson, *Merchant Prince of Boston*, p. 372.
111. THP, Boston, to John P. Cushing, January 11, 1827, Perkins Papers, MHS. Also quoted in Seaburg and Paterson, *Merchant Prince of Boston*, pp. 340–1.
112. J. Downs, 'Fair Game', pp. 146–7; Forbes, *Personal Reminiscences*, pp. 188–96.
113. John Lauritz Larson, *Bonds of Enterprise: John Murray Forbes and Western Development in America's Railway Age* (Cambridge, MA: Harvard University Press, 1984), pp. 85–6 and 89; Paul Goodman, 'Ethics and Enterprise: The Values of a Boston Elite, 1800–1860' in *American Quarterly*, 18, 1966, p. 448.
114. Larson, *Bonds of Enterprise*, pp. 99–100.
115. J. Downs, 'American Merchants and the China Opium Trade, 1800–1840' in *Business History Review*, Winter 1968, p. 438.
116. Cary, *The Memoir of Thomas Handasyd Perkins*, p. 222.

117. Goldstein, *Philadelphia and the China Trade*, pp. 42–3.
118. Larson, *Bonds of Enterprise*, p. 169.
119. Ibid., p. 81.
120. Downs, 'Fair Game', p. 141.
121. T.P. Thornton, *Cultivating Gentlemen: The Meaning of Country Life among the Boston Elite, 1785–1860* (New Haven: Yale University Press 1989), pp. 142 and 205.
122. T.S. Arthur, 'Marrying a Merchant' in *Godey's Lady's Book*, October 1842, pp. 160–4. In 'Retiring from Business', Arthur emphasizes the importance of businessmen working 'to correct evils in society' (*Godey's Lady's Book*, June 1843, p. 288).
123. Thornton, *Cultivating Gentlemen*, p. 162.
124. Goldstein, *Philadelphia and the China Trade*, p. 52.
125. Downs, 'Fair Game', p. 141.
126. Downs, 'Commercial Origins', p. 57.
127. Downs, 'Fair Game', p. 135.
128. Ibid., pp. 142–3.
129. 'Medical Temperance Meeting' in *New-York Evangelist*, March 7, 1840. The speech also appeared in *Christian Secretary*, March 13, 1840.
130. Cary, *The Memoir of Thomas Handasyd Perkins*, p. 223.
131. 'Premium Essay on the Opium Trade' in *Boston Medical and Surgical Journal*, February 13, 1839, p. 18.

Index